W0042191

Signalling Across Space Without Wires

*Being a Description of the Work
of Hertz and his Successors*

OLIVER LODGE

CAMBRIDGE UNIVERSITY PRESS

Cambridge, New York, Melbourne, Madrid, Cape Town,
Singapore, São Paolo, Delhi, Mexico City

Published in the United States of America by Cambridge University Press, New York

www.cambridge.org
Information on this title: www.cambridge.org/9781108052122

© in this compilation Cambridge University Press 2013

This edition first published 1900
This digitally printed version 2013

ISBN 978-1-108-05212-2 Paperback

CAMBRIDGE LIBRARY COLLECTION

Books of enduring scholarly value

Technology

The focus of this series is engineering, broadly construed. It covers technological innovation from a range of periods and cultures, but centres on the
technological achievements of the industrial era in the West, particularly
in the nineteenth century, as understood by their contemporaries. Infra-
structure is one major focus, covering the building of railways and canals,
bridges and tunnels, land drainage, the laying of submarine cables, and the
construction of docks and lighthouses. Other key topics include developments in industrial and manufacturing fields such as mining technology,
the production of iron and steel, the use of steam power, and chemical
processes such as photography and textile dyes.

Signalling Across Space Without Wires

The early 1890s saw the development of wireless telegraphy. Although the
behaviour of radio waves had been predicted by James Clerk Maxwell, the
production of a working coherer occupied some of the greatest practical
physicists of the time. A giant in the field was Heinrich Hertz (1857–94), who
was among the first to discover that radio waves could travel independently
of wires. When Hertz died, his work was continued and soon led to the
development of the first wireless radios. This book, published in 1900, is
the third edition of Sir Oliver Lodge's popular explanation of Hertz's work.
Including the Royal Institution lecture that Lodge (1851–1940) gave in 1894,
along with detailed diagrams, it covers the basic principles of radio waves
and some of the theory surrounding telegraphic technology. Also included in
this reissue is Lodge's 1924 lecture on electrical precipitation, discussing the
scintillating possibility of altering atmospheric conditions through the use
of electrical charges.

Cambridge University Press has long been a pioneer in the reissuing of out-of-print titles from its own backlist, producing digital reprints of books that are still sought after by scholars and students but could not be reprinted economically using traditional technology. The Cambridge Library Collection extends this activity to a wider range of books which are still of importance to researchers and professionals, either for the source material they contain, or as landmarks in the history of their academic discipline.

Drawing from the world-renowned collections in the Cambridge University Library and other partner libraries, and guided by the advice of experts in each subject area, Cambridge University Press is using state-of-the-art scanning machines in its own Printing House to capture the content of each book selected for inclusion. The files are processed to give a consistently clear, crisp image, and the books finished to the high quality standard for which the Press is recognised around the world. The latest print-on-demand technology ensures that the books will remain available indefinitely, and that orders for single or multiple copies can quickly be supplied.

The Cambridge Library Collection brings back to life books of enduring scholarly value (including out-of-copyright works originally issued by other publishers) across a wide range of disciplines in the humanities and social sciences and in science and technology.

Yours truly

H. Hertz

SIGNALLING ACROSS SPACE

WITHOUT WIRES.

BEING A DESCRIPTION OF

THE WORK OF HERTZ & HIS SUCCESSORS.

BY

PROF. OLIVER J. LODGE, F.R.S.

THIRD EDITION,

With Additional Remarks concerning the Application to Telegraphy, and Later Developments.

LONDON:
"THE ELECTRICIAN" PRINTING AND PUBLISHING COMPANY, LIMITED,
SALISBURY COURT, FLEET STREET.

TABLE OF CONTENTS.

———◆———

SIGNALLING THROUGH SPACE WITHOUT WIRES.

THE WORK OF HERTZ

AND

SOME OF HIS SUCCESSORS.

The following pages (up to page 42) are the Notes of a Lecture delivered by Dr. O. J. Lodge before the Royal Institution of Great Britain on Friday evening, June 1, 1894. These notes have been revised by Dr. Lodge, and prepared for publication in the form here presented. After page 42 an account is given of the later applications of Hertzian wave experiments to wireless telegraphy, and a series of Appendices are also given.

Introductory.—1894.

THE untimely end of a young and brilliant career cannot fail to strike a note of sadness and awaken a chord of sympathy in the hearts of his friends and fellow-workers. Of men thus cut down in the early prime of their powers there will occur to us here the names of Fresnel, of Carnot, of Clifford, and now of Hertz. His was a strenuous and favoured youth; he was surrounded from his birth with all the influences that go to make an accomplished man of science—accomplished both on the experimental and on the mathematical side. The front rank of scientific workers is weaker by his death, which occurred on January 1, 1894, the thirty-seventh year of his life. Yet did he not go till he had effected an achievement which will hand his name down to posterity as the founder of an epoch in experimental physics.

In mathematical and speculative physics others had sown the seed. It was sown by Faraday, it was sown by Thomson and by Stokes, by Weber also doubtless, and by Helmholtz; but in this particular department it was sown by none more fruitfully and plentifully than by Clerk Maxwell. Of the seed thus sown Hertz reaped the fruits. Through his experimental discovery, Germany awoke to the truth of Clerk Maxwell's theory of light, of light and electricity combined, and the able army of workers in that country (not forgetting some in Switzerland, France, and Ireland) have done most of the gleaning after Hertz.

This is the work of Hertz which is best known, the work which brought him immediate fame. It is not always that public notice is so well justified. The popular instinct is generous and trustful, and it is apt to be misled. The scientific eminence accorded to a few energetic persons by popular estimate is more or less amusing to those working on the same lines. In the case of Hertz no such mistake has been made. His name is not over well-known, and his work is immensely greater in every way than that of several who have made more noise.

His best known discovery is by no means his only one, and no less than eighteen Papers were contributed to German periodicals by him, in addition to the papers incorporated in his now well-known book on electric waves.

In closing these introductory and personal remarks, I should like to say that the enthusiastic admiration for Hertz's spirit and character felt and expressed by students and workers who came into contact with him is not easily to be exaggerated. Never was a man more painfully anxious to avoid wounding the susceptibilities of others; and he was accustomed to deprecate the prominence given to him by speakers and writers in this country, lest it might seem to exalt him unduly above other and older workers among his own sensitive countrymen.

Speaking of the other great workers in physics in Germany, it is not out of place to record the sorrow with which we have heard of the recent death of Dr. August Kundt, Professor in the University of Berlin, successor to Von Helmholtz in that capacity.

When I consented to discourse on the work of Hertz, my intention was to repeat some of his actual experiments, and especially to demonstrate his less-known discoveries and observations. But the fascination exerted upon me by electric oscillation experiments, when I, too, was independently working at them in the spring of 1888,* resumed its hold, and my lecture will accordingly consist of experimental demonstrations of the outcome of Hertz's work rather than any precise repetition of portions of that work itself.

In case a minority of my audience are in the predicament of not knowing anything about the subject, a five minutes' explanatory prelude may be permitted ; and the simplest way will be for me hastily to summarise our knowledge of the subject before the era of Hertz.

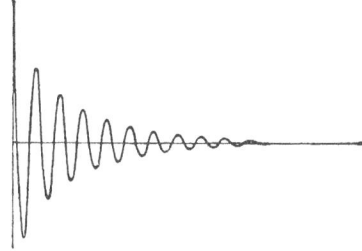

FIG. 1.—Oscillations of Dumb-bell Hertz Vibrator (after Bjerknes).

Just as a pebble thrown into a pond excites surface ripples, which can heave up and down floating straws under which they pass, so a struck bell or tuning fork emits energy into the air in the form of what are called sound waves, and this radiant energy is able to set up vibrations in other suitable elastic bodies.

If the body receiving them has its natural or free vibrations violently damped, so that when left to itself it speedily returns to rest (Fig. 1), then it can respond fully to notes of almost any pitch. This is the case with your ears and the tones of my voice. Tones must be exceedingly shrill before they cease to excite the ear at all.

---

* *Phil. Mag.*, XXVI., pp. 229, 230, August, 1888 ; or "Lightning Conductors and Lightning Guards," pp. 104, 105 ; also *Proc. Roy. Soc.*, Vol. 50, p. 27.

If, on the other hand, the receiving body has a persistent period of vibration, continuing in motion long after it is left to itself (Fig. 2) like another tuning-fork or bell, for instance, then far more facility of response exists, but great accuracy of tuning is necessary if it is to be fully called out; for if the receiver is not thus accurately syntonised with the source, it fails more or less completely to resound.

Conversely, if the *source* is a persistent vibrator, correct tuning is essential, or it will destroy at one moment (Fig. 3)

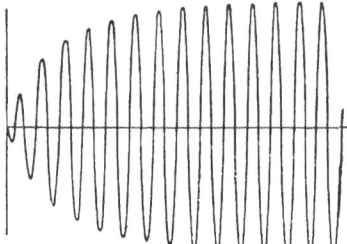

FIG. 2.—Oscillation of Ring-shaped Hertz Resonator excited by Syntonic Vibrator (after Bjerknes).

motion which it originated the previous moment. Whereas, if it is a dead-beat or strongly-damped exciter, almost anything will respond equally well or equally ill to it.

What I have said of sounding bodies is true of all vibrators in a medium competent to transmit waves. Now a sending telephone or a microphone, when spoken to, emits waves into

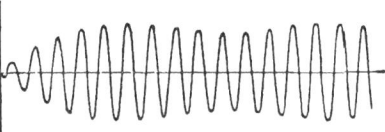

FIG. 3.—Oscillation of Ring Resonator not quite syntonic with Radiator. (For method of obtaining these curves see Fig. 14.)

the ether, and this radiant energy is likewise able to set up vibration in suitable bodies. But we have no delicate means of directly detecting these electrical or etherial waves; and if they are to produce a perceptible effect at a distance, they must be confined, as by a speaking-tube, prevented from spreading, and concentrated on the distant receiver.

This is the function of the telegraph wire; it is to the ether what a speaking-tube is to air. A metal wire in air (*in function, not in details of analogy*) is like a long hollow cavity surrounded by nearly rigid but slightly elastic walls.

Sphere charged from Electrophorus.

Furthermore, any conductor electrically charged or discharged with sufficient suddenness must emit electrical waves into the ether, because the charge given to it will not settle down instantly, but will surge to and fro several times first; and these surgings or electric oscillations must, according to Maxwell, start waves in the ether, because at the end of each half-swing they cause electrostatic, and at the middle of each half-swing they cause electromagnetic effects, and the rapid alternation from one of these modes of energy to the other constitutes etherial waves.* If a wire is handy they will run along it, and may be felt a long way off. If no wire exists they will spread out like sound from a bell, or light from a spark, and their intensity will decrease according to the inverse square of the distance.

Maxwell and his followers well knew that there would be such waves; they knew the rate at which they would go, they knew that they would go slower in glass and water than in air, they knew that they would curl round sharp edges, that they would be partly absorbed but mainly reflected by conductors, that if turned back upon themselves they would produce the phenomena of stationary waves, or interference, or nodes and loops; it was known how to calculate the length of such waves, and even how to produce them of any required or predetermined wave length from 1,000 miles to a foot. Other things were known about them which would take too long to enumerate; any homogeneous insulator would transmit them, would refract or concentrate them if it were of suitable shape, would reflect none of a particular mode of vibration at a certain angle, and so on, and so on.

---

* Strictly speaking, in the waves themselves there is no lag or difference of phase between the electric and the magnetic vibrations; the difference exists in emitter or absorber, but not in the transmitting medium. True radiation of energy does not begin till about a quarter wave-length from the source, and within that distance the initial quarter period difference of phase is obliterated.

All this was *known*, I say, known with varying degrees of confidence; but by some known with as great confidence as, perhaps even more confidence than, is legitimate before the actuality of experimental verification.

Hertz supplied the verification. He inserted suitable conductors in the path of such waves, conductors adapted for the occurrence in them of induced electric oscillations, and to the surprise of everyone, himself doubtless included, he found that the secondary electric surgings thus excited were strong enough to disproy themselves by minute electric sparks.

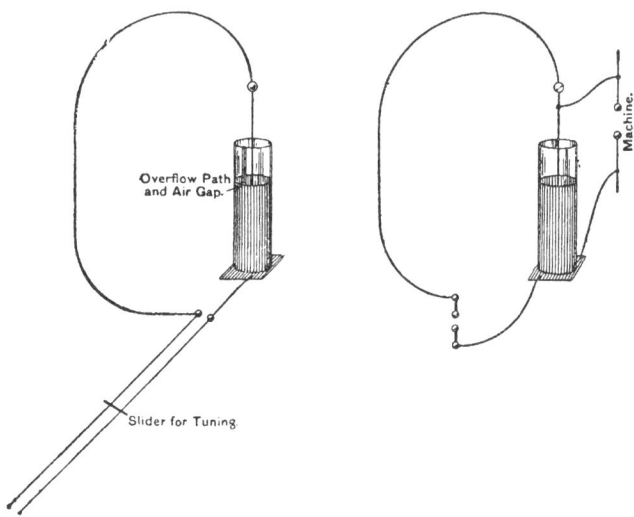

Fig. 4.—Experiment with Syntonic Leyden Jars (*cf.* page 21).

Syntonic Leyden Jars.

I shall show this in a form which requires great precision of tuning or syntony, both emitter and receiver being persistently vibrating things giving some 30 or 40 swings before damping has a serious effect. I take two Leyden jars with circuits about a yard in diameter, and situated about two yards apart (Fig. 4). I charge and discharge one jar, and observe that the surgings set up in the other can cause it to overflow if it is syntonised with the first.[*]

[*] See *Nature*, Vol. XLI., p. 368, where I first described this experiment; or quotation in J. J. Thomson's " Recent Researches," p. 395.

A closed circuit such as this is a feeble radiator and a feeble absorber, so it is not adapted for action at a distance. In fact, I doubt whether it will visibly act at a range beyond the $\frac{1}{4}\lambda$ at which true radiation of broken-off energy occurs. If the coatings of the jar are separated to a greater distance, so that the dielectric is more exposed, it radiates better ; because in true radiation the electrostatic and the magnetic energies must be equal, whereas in a ring circuit the magnetic energy predominates. By separating the coats of the jar as far as possible we get a typical Hertz vibrator (Fig. 5), whose dielectric extends out into the room, and thus radiates very powerfully.

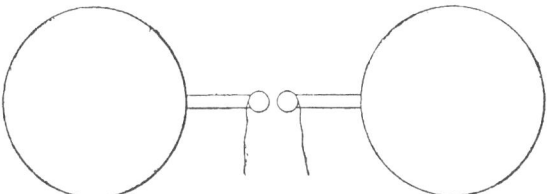

Fig. 5.—Standard Hertz Radiator.

Ordinary Size Hertz Vibrator.

In consequence of its radiation of energy, its vibrations are rapidly damped, and it only gives some three or four good strong swings (Fig. 1). Hence it follows that it has a wide range of excitation ; *i.e.*, it can excite sparks in conductors barely at all in tune with it.

The two conditions, conspicuous energy of radiation and persistent vibration electrically produced, are at present incompatible. Whenever these two conditions coexist, considerable power or activity will, of course, be necessary in the source of energy. At present they only coexist in the sun and other stars, in the electric arc, and in furnaces.

Two Circular Vibrators Sparking in sympathy.

The receiver Hertz used was chiefly a circular resonator (Fig. 6), not a good absorber but a persistent vibrator, well adapted for picking up disturbances of precise and measurable wave-length. Its mode of vibration when excited by emitter in tune with it is depicted in Fig. 2. I find that the circular resonators can act as senders too ; here is one exciting quite long sparks in a second one.

Electric Syntony.—That was his discovery, but he did not stop there. He at once proceeded to apply his discovery to the verification of what had already been predicted about the waves, and by laborious and difficult interference experiments he ascertained that the previously calculated length of the waves was thoroughly borne out by fact. These interference experiments in free space are his greatest achievement.

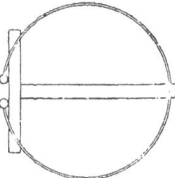

FIG. 6.—Circular Resonator. (The knobs ought to nearly touch each other.)

He worked out every detail of the theory splendidly, separately analysing the electric and the magnetic oscillation, using language not always such as we should use now, but himself growing in theoretic insight through the medium of

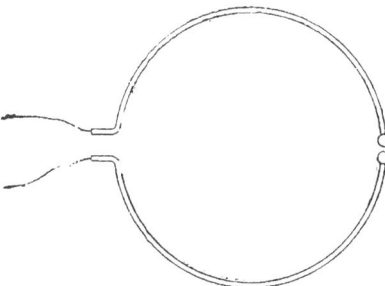

FIG. 6A.—Any circular Resonator can be used as a sender by bringing its knobs near the sparking knobs of a coil ; but a simple arrangement is to take two semi-circles, as in above figure, and make them the coil terminals. The capacity of the cut ends can be varied, and the period thereby lengthened, by expanding them into plates.

what would have been to most physicists a confusing maze of troublesome facts, and disentangling all their main relations most harmoniously.

Holtz Machine, A and B Sparks : Glass and Quartz Panes in Screen.

While Hertz was observing sparks such as these, the primary or exciting spark and the secondary or excited one, he observed as a bye issue that the secondary spark occurred more easily if the light from the primary fell upon its knobs. He examined this new influence of light in many ways, and showed that although spark light and electric brush light were peculiarly effective, any source of light that gave very ultra-violet rays produced the same result.

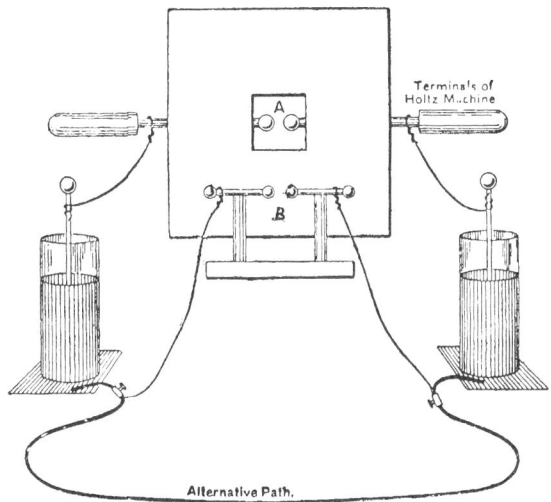

Fɪɢ. 7. —Experiment arranged to show effect on one spark of light from another. The **B** spark occurs more easily when it can see the **A** spark through the window, unless the window is glazed with glass. A quartz pane transmits the effect : glass cuts it off.

The above figure represents my way of showing the experiment. It will be observed that with this arrangement the **B** knobs are at the same potential up to the instant of the flash, and in that case the ultra-violet portion of the light of the **A** spark assists the occurrence of the **B** spark. But it is interesting to note what Elster and Geitel have found (*see* Appendix IV., Fig. 59), that if the **B** knobs were subjected to steady strain instead of to impulsive rush—*e.g.*, if they were connected to the inner coats of the jars instead of the outer coatings—that then the effect of ultra-violet light on either spark-gap would exert a deterrent influence, so that the spark would probably occur at the other, or non-illuminated gap. With the altered connections it is, of course, not feasible to illuminate one spark by the light of the other; the sparks are then alternative, not successive.

Wiedemann and Ebert, and a number of experimenters, have repeated and extended this discovery, proving that it is the cathode knob on which illumination takes effect; and Hallwachs and Righi made the important observation, which Elster and Geitel, Stoletow, Branly, and others have extended, that

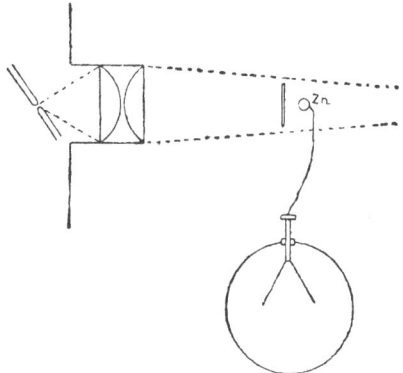

Fig. 8.—Zinc Rod in Arc Light, protected by Glass Screen. The lenses are of quartz, but there is no need for any lenses in this experiment; leakage of electricity begins directly the glass plate is withdrawn.

a freshly-polished zinc or other oxidisable surface, if charged negatively, is gradually discharged by ultra-violet light.

It is easy to fail in reproducing this experimental result if the right conditions are not satisfied; but if they are it is absurdly easy, and the thing might have been observed nearly a century ago.

Zinc discharging Negative Electricity in Light ; Gold Leaf Electroscope ; Glass and Quartz Panes ; Quartz Prism.

Take a piece of zinc, clean it with emery paper, connect it to a gold leaf electroscope, and expose it to an arc lamp. (Fig. 8).

If charged positively nothing appears to happen, the action is very slow ; but a negative charge leaks away in a few seconds if the light is bright. Any source of light rich in ultra-violet rays will do ; the light from a spark is perhaps most powerful of all. A pane of glass cuts off all the action ; so does atmospheric air in sufficient thickness (at any rate, town air), hence sunlight is not powerful. A pane of quartz transmits the action almost undiminished, but fluorspar may be more transparent still. Condensing the arc rays with a quartz lens and analysing them with a quartz prism or reflection grating, we find that the most effective part of the light is high up in the ultra-violet, surprisingly far beyond the limits of the visible spectrum* (Fig. 9, next page).

This is rather a digression, but I have taken some pains to show it properly because of the interest betrayed by Lord

---

* While preparing for the lecture it occurred to me to try, if possible during the lecture itself, some new experiments on the effect of light on negatively charged bits of rock and ice, because if the effect is not limited to metals it must be important in connection with atmospheric electricity. When Mr. Branly coated an aluminium plate with an insulating varnish, he found that its charge was able to soak in and out of the varnish during illumination (*Comptes Rendus*, Vol. CX., p. 898, 1890). Now the mountain tops of a negatively charged earth are exposed to very ultra-violet rays, and the air is a dielectric in which quiet up-carrying and sudden downpour of electricity could go on in a manner not very unlike the well-known behaviour of water vapour ; and this perhaps may be the reason, or one of the reasons, why it is not unusual to experience a thunderstorm after a few fine days. I have now tried these experiments on such geological fragments as were handy, and find that many of them discharge negative electricity under the action of a naked arc, especially from the side of the specimens which was somewhat dusty, but that when wet they discharge much less rapidly, and when positively charged hardly at all. Ice and garden soil discharge negative electrification, too, under ultra-violet illumination, but not so quickly as limestone, mica schist, ferruginous quartz, clay, and some other specimens. Granite barely acts ; it seems to insulate too well. The ice and soil were tried in their usual moist condition, but, when thoroughly dry, soil discharges quite rapidly. No rock tested was found to discharge as quickly as does a surface of perfectly bright metal, such as iron, but many discharged much more quickly than ordinary dull iron, and rather more quickly than when the bright iron surface was thinly oiled or wetted with water. To-day (June 5, 1894) I find that the leaves of Geranium discharge positive electrification five times as quickly as negative, under the action of an arc-light, and that glass cuts the effect off while quartz transmits it. [For Elster and Geitel's experiments, and those of Righi, *see* Appendices, p. 115 *et seq.*]

Kelvin on this matter, and the caution which he felt about accepting the results of the Continental experimenters too hastily.

It is probably a chemical phenomenon, and I am disposed to express it as a modification of the Volta contact effect* with illumination.

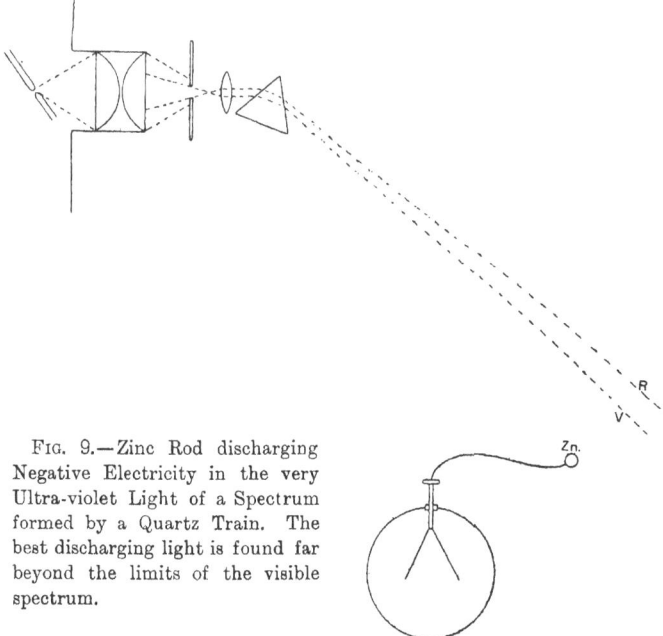

FIG. 9.—Zinc Rod discharging Negative Electricity in the very Ultra-violet Light of a Spectrum formed by a Quartz Train. The best discharging light is found far beyond the limits of the visible spectrum.

Return now to the Hertz vibrator, or Leyden jar with its coatings well separated, so that we can get into its electric as well as its magnetic field. Here is a great one giving waves 30 metres long, radiating while it lasts with an activity of 100 H.P., and making ten million complete electric vibrations per second. It is made of four large copper sheets soldered together two and two, strung up by thin rope to a gallery, and each pair connected with the other by several yards of No. 0 pure copper rod, interrupted by a pair of sparking knobs (Fig. 10).

* *See* B. A. Report, 1884, pp. 502-519 ; or *Phil. Mag.*, Vol. XIX., pp. 267-352·

Large Hertz Vibrator in action ; Abel's Fuse Detector ; Vacuum Tube Detector ; Striking of an Arc.

Its great radiating power damps it down very rapidly, so that it does not make above two or three swings ; but nevertheless, each time it is excited, sparks can be drawn from most of the reasonably elongated conductors in the theatre of this Institution, and indeed from wire fencing and iron roofs outside this building.

A suitably situated gas-leak can be ignited by these induced sparks. An Abel's fuse connecting the water pipes with the gas pipes will blow off; vacuum tubes connected to nothing will glow (this fact has been familiar to all who have worked with Hertz waves since 1889), electric leads, if anywhere near each other, as they are in some incandescent lamp-holders, may spark across to each other, thus striking an arc and blowing their fuses. This blowing of fuses by electric radiation frequently happened at Liverpool till the suspensions of the theatre lamps were altered. They had at first been held in position by wire guides, which served as collectors of the Hertz waves or impulses.

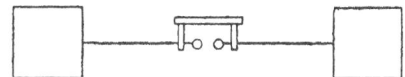

Fig. 10—Hertz Oscillator on reduced scale, $\frac{1}{10}$th inch to a foot.

The striking of an arc by the little reverberating sparks between two lamp-carbons connected with the 100-volt mains I incidentally now demonstate. An arc is started directly the large Hertz vibrator is exicited at a distance.

There are some who think that lightning flashes can do none of these secondary things. They are mistaken.

Specimens of Emitters and Receivers.

On the table are specimens of various emitters and receivers such as have been used by different people ; the orthodox Hertz radiator (Fig. 5), and the orthodox Hertz receivers :— A circular ring (Fig. 6) for interference experiments, because it is but little damped, and a straight wire for receiving at a distance, because it is a much better absorber. Beside these are the spheres and ellipsoids (or elliptical plates), which I have myself introduced and mainly used (Fig. 19), because they are powerful radiators and absorbers, and because their theory has been worked out by Horace Lamb and J. J. Thomson. Also dumb-bells (Fig. 11) without air-gap, which must be excited by a positive spark at one end and a negative spark at the other, and many other shapes, the most recent of mine being the inside of a hollow cylinder with sparks at ends of a diameter (Fig. 12); this being a feeble radiator, but a very persistent vibrator,* and, therefore, well adapted for interferance and diffraction experiments. But, indeed, spheres can be made to vibrate longer than usual by putting them into copper hats or enclosures, in which an aperture of varying size can be made to let the waves out (Figs 20 and 21).

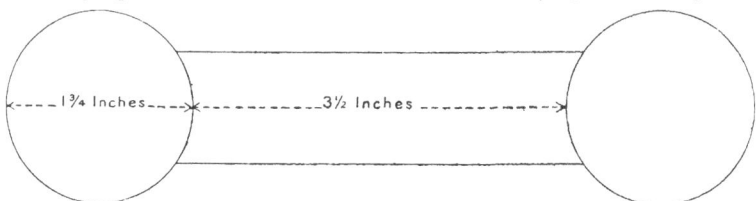

FIG. 11.—A Small Dumb-bell Form of Radiator for Impulsive Rush.

Many of these senders will do for receivers too, giving off sparks to other insulated bodies or to earth ; but besides the Hertz type of receiver, many other detectors of radiation have

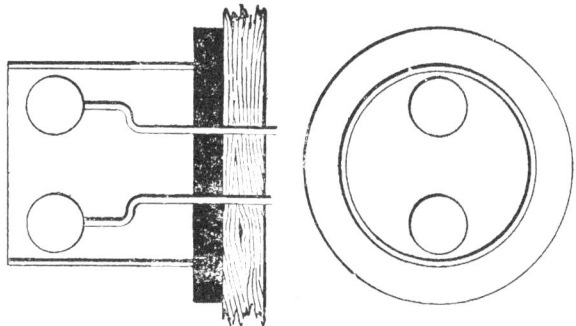

FIG. 12.—Dr. Lodge's Hollow Cylindrical Radiator, arranged horizontally against the outside of a Metal-lined Box containing the Spark-producing Apparatus. Half natural size. Emitting 3in. waves.

been employed. Vacuum tubes can be used, either directly or on the trigger principle, as by Zehnder (Fig. 13),† the resonator spark precipitating a discharge from some auxiliary

---

* J. J. Thomson, "Recent Researches," 344.   † *Wied. Ann.*, XLVII., p. 77·

battery or source of energy, and so making a feeble disturbance very visible. Explosives may be used for the same purpose, either in the form of mixed water-gases or in the form of an Abel's fuse. FitzGerald found that a tremendously sensitive galvanometer could indicate that a feeble spark had passed, by reason of the consequent disturbance of electrical equilibrium which settled down again through the galvanometer.* This was the method he used in this theatre four years ago. Blyth used a one sided electrometer, and V. Bjerknes has greatly developed this method (Fig. 14), abolishing the need for a spark, and making the electrometer metrical, integrating and satisfactory.† With this detector many measurements have been made at Bonn by Bjerknes, Yule, Barton and others on waves concentrated and kept from space dissipation by guiding wires.

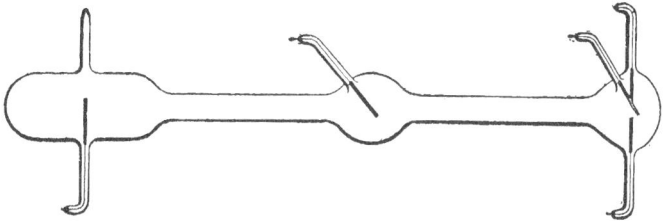

Fig. 13.—Zehnder's Trigger Tube. Half Natural Size. The two right-hand terminals, close together, are attached to the Hertz receiver ; another pair of terminals are connected to some source just not able to make the tube glow until the scintilla occurs and makes the gas more conducting— as observed by Schuster and others.

Mr. Boys has experimented on the mechanical force exerted by electrical surgings, and Hertz also made observations of the same kind.

Various Detectors.

Going back to older methods of detecting electrical radiation, we have, most important of all, a discovery made long before man existed, by a creature that developed a sensitive cavity on its skin ; a creature which never so much as had a name to be remembered by (though perhaps we now call it trilobite). Then, in recent times we recall the photographic

* FitzGerald, *Nature*, Vol. XLI., p. 295, and Vol. XLII., p. 172.
† *Wied. Ann.*, 44, p. 74.

plate and the thermopile, with its modification, the radio-micrometer ; also the so-called bolometer, or otherwise-known Siemens' pyrometer, applied to astronomy by Langley, and applied to the detection of electric waves in wires by Rubens and Ritter and Paalzow and Arons. The thermal junction was applied to the same purpose by Kolacek, D. E. Jones and others.

And, before all these, the late Mr. Gregory, of Cooper's Hill, made his singularly sensitive expansion meter, whereby waves in free space could be detected by the minute rise of temperature they caused in a platinum wire, a kind of early and sensitive form of Cardew voltmeter.

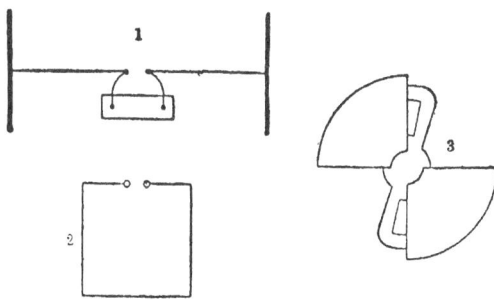

Fig. 14.—Bjerknes' Apparatus, showing (1) a Hertz vibrator connected to an induction coil ; (2) a nearly closed circuit receiver properly tuned with the vibrator ; and (3) a one-sided electrometer for inserting in the air-gap of 2. The receiver is not provided with knobs, as shown, but its open circuit is terminated by the quadrants of the electrometer, which is shown on an enlarged scale alongside. The needle is at zero potential and is attracted by both quadrants. By calculation from the indications of this electrometer Bjerknes plotted the curves 1, 2 and 3 on pages 4 and 5. Fig. 1 represents the oscillations of the primary vibrator, rapidly damped by radiation of energy. Fig. 2 represents the vibrations thereby set up in the resonating circuit when the two are accurately in tune ; and which persist for many swings. Fig. 3 shows the vibrations excited in the same circuit when slightly out of tune with the exciter. A receiver of this kind makes many swings before it is seriously damped, though the open plate radiator does not.

Going back to the physiological method of detecting surgings, Hertz tried the frog's-leg nerve-muscle preparation, which to the steadier types of electrical stimulus is so sur-passingly sensitive, and to which we owe the discovery of current electricity. But he failed to get any result. Ritter

has succeeded; but, in my experience, failure is the normal
and proper result. Working with my colleague, Prof. Gotch,
at Liverpool, I too have tried the nerve and muscle preparation
of the frog (Fig. 15), and we find that an excessively violent
stimulus of a rapidly alternating character, if pure and
unaccompanied by secondary actions, produces no effect—
no stimulating effect, that is—even though the voltage is so
high that sparks are ready to jump between the needles in
direct contact with the nerve.

All that such oscillations do, if continued, is to produce a
temporary paralysis or fatigue of the nerve, so that it is unable
to transmit the nerve impulses evoked by other stimuli, from
which paralysis it recovers readily enough in course of time.

This has been expected from experiments on human beings,
such experiments as Tesla's and those of d'Arsonval. But an
entire animal is not at all a satisfactory instrument wherewith
to attack the question; its nerves are so embedded in con-

Fig. 15.—Experiment of Gotch and Lodge on the physiological effect of
rapid pure electric alternations. Nerve-muscle preparation, with four
needles, or else non-polarisable electrodes applied to the nerve. C and D
are the terminals of a rapidly-alternating electric current from a conductor
at zero potential (namely, the terminals of a derived circuit from a wire
connecting the outer coats of a pair of discharging Leyden jars), while
A and B are the terminals of an ordinary very weak galvanic or induction
coil stimulus only just sufficient to make the muscle twitch. The C D
terminals do not stimulate the nerve, though at very high alternative
potentials, but they gradually and temporarily paralyse it, so that the test
terminals A B produce no effect for a time.

ducting tissues that it may easily be doubted whether the
alternating type of stimulus ever reaches them at all. By
dissecting out a nerve and muscle from a deceased frog after
the historic manner of physiologists, and applying the
stimulus direct to the nerve, at the same time as some other
well known $\frac{1}{100}$th of a volt stimulus is applied to another
part of the same nerve further from the muscle, it can be
shown that rapid electric alternations, if entirely unaccompanied by static charge or by resultant algebraic electric
transmission, evoke no excitatory response until they are so
violent as to give rise to secondary effects such as heat or
mechanical shock. Yet, notwithstanding this inaction, they

c

gradually and slowly exert a paralysing or obstructive action on the portion of the nerve to which they are applied, so that the nerve impulse excited by the feeble just perceptible $\frac{1}{100}$th-volts stimulus above is gradually throttled on its way down to the muscle, and remains so throttled for a time varying from a few minutes to an hour after the cessation of the violence. [I did not show this experiment at the lecture.]

Air Gap and Electroscope charged by Glass Rod and discharged by the Wave Impulse from a moderately distant Sphere excited by Coil.

Among trigger methods of detecting electric radiation, I have spoken of the Zehnder vacuum tubes ; another method is one used by Boltzmann.* A pile of several hundred volts

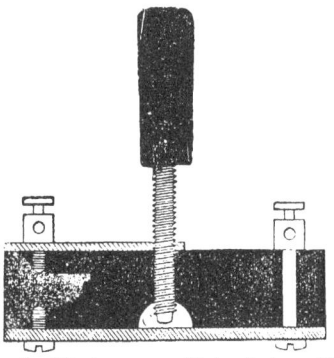

Fig. 16.—Air gap for Electroscope. Natural size. The bottom plate is connected to, and represents, the cap of an electroscope ; the " knob " above it, mentioned in text, is the polished end of the screw, whose terminal is connected with the case of the instrument or " earth." The electroscope being charged to the verge of overflow, the impact of weak electric waves collected by a bit of wire sticking up from the left-hand binding screw precipitates the collapse of the leaves.

is on the verge of charging an electroscope through an air gap just too wide to break down. Very slight electric surgings precipitate the discharge across the gap, and the leaves diverge. I show this in a modified and simple form. On the cap of an electroscope is placed a highly polished knob or rounded end connected to the sole, and just not touching the cap, or, rather, just not touching a plate connected with the cap (Fig. 16), the distance between knob and plate being

---

* *Weid. Ann.*, 40, p. 399,

almost infinitesimal, such a distance as is appreciated in spherometry. Such an electroscope overflows suddenly and completely with any gentle rise of potential. Bring excited glass near it, the leaves diverge gradually and then suddenly collapse, because the air space snaps : remove the glass, and they rediverge with negative electricity; the knob above the cap being then charged positively, and to the verge of sparking. In this condition any electrical waves, collected if weak by a foot or so of wire projecting from the cap, will discharge the electroscope by exciting surgings in the wire, and so breaking down the air gap. The chief interest about this experiment seems to me the extremely definite dielectric strength of so infinitesimal an air space. Moreover, it is a detector for Hertz waves that might have been used last century; it might have been used by Benjamin Franklin.

For to excite them no coil or anything complicated is necessary ; it is sufficient to flick a metal sphere or cylinder with a silk handkerchief and then discharge it with a well-polished knob. If it is not well polished the discharge is comparatively gradual, and the vibrations are weak ; the more polished are the sides of an air gap, the more sudden is the collapse and the more vigorous the consequent radiation, especially the radiation of high frequency, the higher harmonics of the disturbance.

For delicate experiments it is sometimes well to repolish the knobs every hour or so. For metrical experiments it is often better to let the knobs get into a less efficient but more permanent state. This is true of all senders or radiators. For the generation of the, so to speak, " infra-red " long-period Hertz waves any knobs will do, but to generate the " ultra-violet " short-period waves high polish is essential.

Microphonic Detectors.

Receivers or detectors, which for the present I temporarily call microphonic, are liable to respond best to the more rapid vibrations. Their sensitiveness is to me surprising, though of course it does not approach the sensitiveness of the eye ; at the same time I am by no means sure that the eye differs from them in kind. It is these detectors that I wish specially to bring to your notice.

Prof. Minchin, whose long and patient work in connection with photo-electricity is now becoming known, and who has devised an instrument more sensitive to radiation than even Boys' radiomicrometer, in that it responds to the radiation of a star while the radiomicrometer does not, found some years ago that some of his light-excitable cells lost their sensitiveness capriciously on tapping, and later he found that they frequently regained it again while Mr. Gregory's Hertz-wave experiments were going on in the same room.

These " impulsion-cells," as he terms them, are troublesome things for ordinary persons to make and work with—at least I have never presumed to try—but in Mr. Minchin's hands they are surprisingly sensitive to electric waves.*

The sensitiveness of selenium to light is known to everyone, and Mr. Shelford Bidwell has made experiments on the variations of conductivity exhibited by a mixture of sulphur and carbon.

Nearly four years ago M. Edouard Branly found that a burnished coat of porphyrised copper spread on glass or ebonite, diminished its resistance enormously, from some millions to some hundreds of ohms, when it was exposed to the neighbourhood, even the distant neighbourhood, of Leyden jar or coil sparks. He likewise found that a tube of metallic filings behaved similarly, and that both recovered their original resistance on shaking or tapping. Mr. Croft exhibited this fact recently at the Physical Society. M. Branly also made pastes and solid rods of filings, in Canada balsam and in sulphur, and found them likewise sensitive.†

With me the matter arose somewhat differently, as an outcome of the air-gap detector employed with an electroscope by Boltzman (Fig. 16). For I had observed in 1889 that two knobs sufficiently close together, far too close to stand any voltage such as an electroscope can show, could, when a spark passed between them, actually cohere ; conducting an ordinary bell-ringing current if a single voltaic cell was in circuit ; and, if there were no such cell, exhibiting an electromotive force of their own sufficient to disturb a low resistance galvanometer

---

* *Phil. Mag.*, Vol. XXXI., p. 223.

† E. Branly, *Comptes Rendus*, Vol. CXI., p. 785 ; and Vol. CXII., p. 90.

vigorously, and sometimes requiring a faintly perceptible amount of force to detach them. The experiment was described to the Institution of Electrical Engineers in 1890,* and Prof. Hughes said he had observed the same thing.

The experiment of the syntonic Leyden jars can be conveniently shown with the double knob or 1889 coherer. The pair of knobs are arranged to connect the coatings of the receiving jar (a large condenser being interposed to prevent their completing a purely metallic circuit), and in circuit with

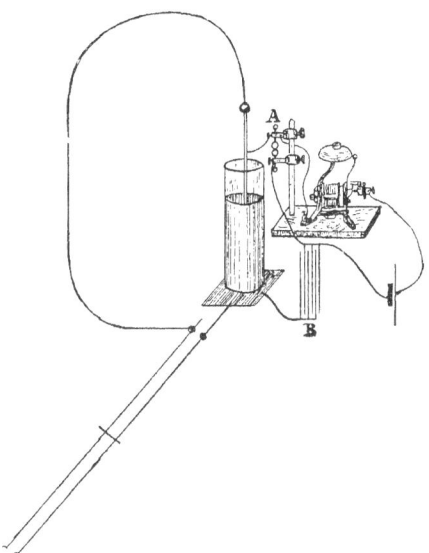

Fig. 16A.—Receiver in Syntonic Jar Experiment, with Knob Coherer and Tapper-back (*cf.* Fig. 4).

them is a battery and a bell. Every time the receiving jar responds syntonically to the electric vibration of the other jar, the knobs cohere (if properly adjusted) and the bell rings. If the bell is free in air it continues ringing until the knobs are gently tapped asunder; but if the bell stands on the same table as the knobs, especially if it rests one foot on the actual stand, then its first stroke taps them back instantly and automatically, and so every discharge of the sending jar is

* *Journal* Institution of Electrical Engineers, 1890, Vol. XIX., pp. 352-4 ; or "Lightning Conductors and Lightning Guards," pp. 382-4.

signalled by a single stroke of the bell.   Here we have in essence a system of very distinctly syntonic telegraphy, for the jars and their circuits must be accurately tuned together, if there is to be any response.   A very little error in tuning, easily made by altering the position of the slider (Fig. 4), will make them quite unresponsive, unless the distance between them is reduced.

At the maximum distance of response the tuning required is excessively sharp.   But, certainly, for these closed and durably-vibrating circuits, the distance of response is small, as has been said before.   Fig. 16A shows the syntonic

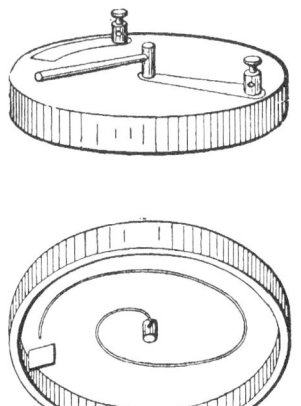

Fig. 17.—Early Form of Coherer, consisting of a spiral of thin iron wire mounted on an adjustable spindle and an aluminium plate.   When the lever is moved clockwise the tip of the iron wire presses gently against the aluminium plate, whose end is bent at right angles and passed through into the hollow circular wooden box, of which the upper figure shows the top and general appearance, and the lower figure shows the inside.

Leyden jar experiment arranged with the double knob coherer, instead of with the spark gap of Fig. 4.

Coherer in open, responding to Feeble Stimuli :—Small Sphere, Gas-lighter, Distant Sphere, Electrophorus.

Well, this arrangement, which I call a coherer, is the most astonishingly sensitive detector of Hertz waves.   It differs from an actual air-gap in that the insulating film is not really insulating; the film breaks down not only much more easily, but also in a less discontinuous and more permanent manner, than an air-gap. Branly's tube of filings, a series of bad contacts, clearly works on the same plan ; and though a tube of filings is by no means so sensitive, yet it is in many respects easier to work with, and except for very feeble stimuli, is more metrical. If the filings used are coarse, say turnings or borings, the tube approximates to a single coherer ; if they are fine, it has a larger range of sensibility. In every case what these receivers feel are sudden jerks of current; smooth sinuous vibrations are ineffective. They seem to me to respond best to waves a few inches long, but doubtless that is determined chiefly by the dimensions of some conductor with which they happen to be associated. (Figs. 17 and 18.)

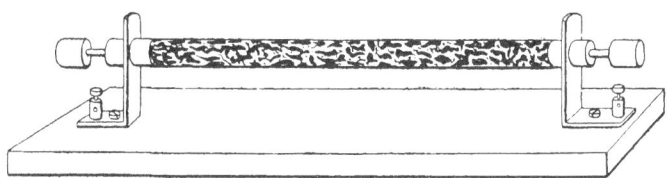

Fɪɢ. 18.—Early Form of Iron Borings Tube. One-half natural size, with solid brass cylinder terminals in each end of the tube, making contact with the borings.

Experiment showing Filings Tube responding to Sphere, to Electrophorus, and to a Quasi-" Spark " from the Discharge of an ordinary Gold-leaf Electroscope.

I picture the action as follows : Suppose two fairly clean pieces of metal in light contact—say two pieces of brass or of iron—connected to a single voltaic cell ; a film of what may be called oxide intervenes between the surfaces so that only an insignificant current is allowed to pass, because a volt or two is insufficient to break down the insulating film, except perhaps at one or two atoms.* If the film is not permitted to conduct at all, it is not very sensitive ; the most sensitive

---

* See *Phil. Mag.*, Jan., 1894, p. 94, where this explanation (whether true or not) was first given, and where the author first published his fuller experience of coherer behaviour.

condition is attained when an infinitesimal current passes, strong enough just to show on a moderate galvanometer.

Now let the slightest surging occur, say by reason of a sphere being charged and discharged at a distance of forty yards ; the film at once breaks down, perhaps not completely —that is a question of intensity—but permanently. As I imagine, more molecules get within each other's range, incipient cohesion sets in and the momentary electric quiver acts somewhat like a flux. It is a singular variety of electric welding. A stronger stimulus enables more molecules to hold on, the process is surprisingly metrical ; and as far as I roughly know at present, the change of resistance is proportional to the energy of the electric radiation, from a source of given frequency.

It is to be specially noted that a battery current is not needed to *effect* the cohesion, only to demonstrate it. The battery can be applied after the spark has occurred, and the resistance will be found changed as much as if the battery had been on all the time.

The incipient cohesion electrically caused can be mechanically destroyed. Sound vibrations, or any other feeble mechanical disturbances, such as scratches or taps, are well adapted to restore the contact to its original high-resistance sensitive condition. The more feeble the electrical disturbance the slighter is the corresponding mechanical stimulus needed for restoration. When working with the radiating sphere (Fig. 19) at a distance of forty yards out of window, I could not for this reason shout to my assistant to cause him to press the key of the coil and make a spark, but I showed him a duster instead, this being a silent signal which had no disturbing effect on the coherer or tube of filings. I mention 40 yards, because that was one of the first outdoor experiments ; but I should think that something more like half-a-mile was nearer the limit of sensitiveness for this particular apparatus as then arranged. However, this is a rash statement not at present verified.* At 40 or 60 yards

---

* This statement has been absurdly misunderstood, as if it was a prediction of what would always be the limit of sensitiveness for any apparatus and any sized sender. Nothing of the kind was in my mind. Such predictions are always preposterous, and I am not obliged to those who imagined that I had been guilty of one of them.—O. J. L., 1899.

the exciting spark could be distinctly heard, and it was interesting to watch the spot of light begin its long excursion and actually travel a distance of 2in. or 3in. before the sound arrived. This experiment proved definitely enough that the efficient cause travelled quicker than sound, and disposed completely of any sceptical doubts as to sound-waves being, perhaps, the real cause of the phenomenon. Signals were obtained across the full width of the college quadrangle, and later, with larger apparatus, between the college tower and another high building half-a-mile away.

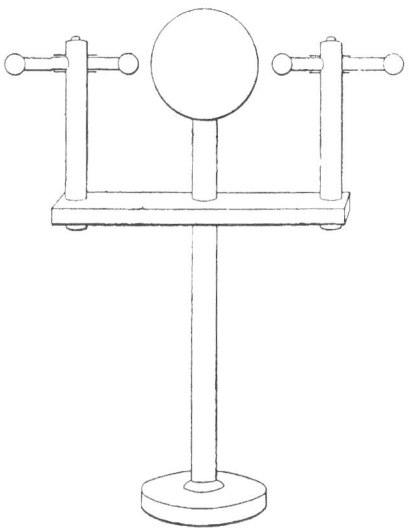

Fig. 19.—Radiator used in the library of the Royal Institution, exciting the Coherer (Fig. 17) on the lecture table in the Theatre. I also used a radiator with two or with three large spheres between two knobs, and described it in *Nature*, Vol. 41, p. 462, 1890. This is the radiator which Prof. Righi has improved and made in a compact form with oil between the two middle spheres.

Invariably, when the receiver is in good condition, sound or other mechanical disturbance acts one way, viz., in the direction of increasing resistance, while electrical radiation or jerks act the other way, decreasing it. While getting the receiver into condition, or when it is getting out of order, vibrations and sometimes electric discharges act irregularly;

and an occasional good shaking does the filings good. I have
taken rough measurements of the resistance by the simple
process of restoring the original galvanometer deflection by
adding or removing resistance coils. A half-inch tube, 8in.
long, of selected iron turnings (Fig. 18) had a resistance of
2,500 ohms in the sensitive state. A feeble stimulus, caused
by a distant electrophorus spark, brought it down 400 ohms.
A rather stronger one reduced it by 500 and 600, while a
trace of spark given to a point of the circuit itself ran it down
1,400 ohms.

This is only to give an idea of the quantities. I have not
yet done any seriously metrical experiments.

Added later.—My assistant, Mr. E. E. Robinson, early
noticed that when a telephone was used as receiver, say with a
single-point coherer (*see* illustration on opposite page), which
is a very sensitive arrangement, every disturbance of the
coherer due to received waves is accompanied by a crackle
or tick in the telephone, without any tapping back being
necessary. This is, indeed, the easiest mode of receiving
signals, and we often practised it. If a suitable, well-damped
galvanometer, such as a Thomson marine speaking-galvano-
meter, is included also in the circuit (a more sensitive one is
sometimes necessary—and we frequently used a D'Arsonval—
but it must be well damped), the meaning of these ticks is
recognised ; each represents a minute change in the resistance
of the coherer—not at all the full change usually employed,
but little subsidiary changes, sometimes up and sometimes
down, barely sufficient to affect a galvanometer, but quite
adequate (being so sudden) to disturb a telephone. This
method of receiving, which at first is very sensitive, after a
time becomes less so ; the point shows signs of fatigue,
probably due to too perfect cohesion having been gradually
established, and a mechanical tap back is desirable to restore
it to its original condition.

If all the signals received were precisely of the same
strength, I doubt if these superposed crimples of resistance
would occur ; but signals depending on quality of sending
spark never are of the same strength, and accordingly the
sudden slight variations of resistance do occur. Usually
an ordinary high-resistance telephone was employed, and it was

Simplest Receiving Arrangement : a Telephone in Circuit with Single-point Coherer without Tapper-back.   B a needle resting against a watch-spring A adjusted by screw C.

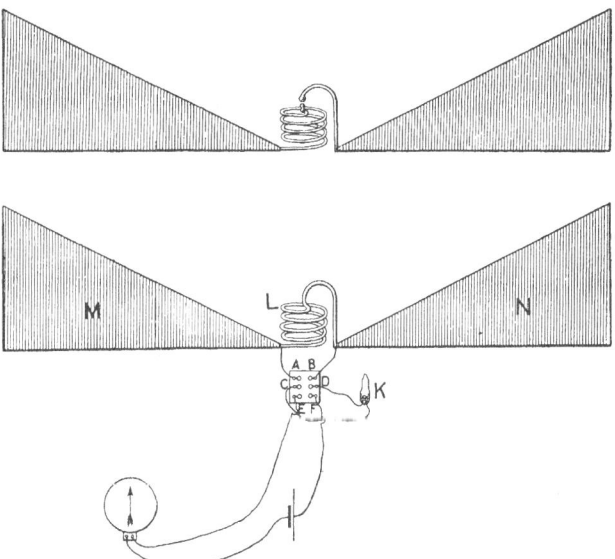

Syntonic Sender and Receiver used in the experiments plotted on page 28.   The switch enables the coherer K to be connected either to the tuned resonator M L N or to the detecting circuit E F.   Weak impulses are felt when the switch is C E, D F ; strong impulses when the switch is C A, D B ; provided the coil L is similar to the coil of the radiator above.   The impulses are plotted in the diagram Fig. 19A.

joined to the coherer circuit through one of the usual small transformers—a plan which has many obvious advantages.

The fluctuations of resistance of a coherer dependent on various strengths of stimulus are instructively shown in some metrical experiments made by Mr. Robinson, and a plotting of which I showed to the Physical Society of London in 1897. This plotting is here reproduced, and it shows the singular fact that, whereas a stronger electrical stimulus usually decreases the resistance, as is natural, a weaker subsequent stimulus usually increases it again : so that alternately

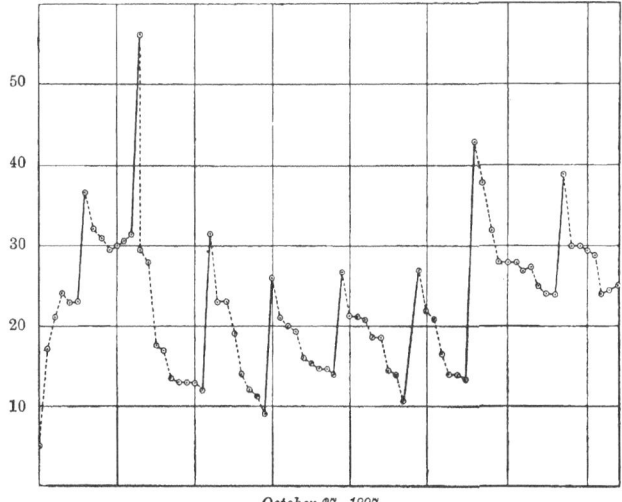

October 27, 1897.

FIG. 19A.—Current through Coherer after successive Electrical Stimuli, without any mechanical tapping back. The sudden rises are obtained when the circuits are syntonised. Weaker stimuli cause the descents.

strong and weak stimuli send the curve zigzagging up and down, until it gets into a condition demanding rejuvenation by a mechanical tap back.

Sometimes a decidedly strong electrical stimulus knocks down the conductivity of the coherer as if it had been tapped back. This is almost certainly due to a burning of the delicate contacts—a blowing of a fuse as it were,—and the effect of this electrical burn back is quite different from the effect of a mechanical tap back, inasmuch as it leaves the coherer insensitive. A shaking up is necessary to restore it.

Continuation of Lecture.

I now call your attention to the Table on next page of various kinds of detector for electric radiation distributed in groups.

Selenium is inserted in this table in the microphone column, because it is a substance which in certain states is well known to behave to visible light as these other microphonic detectors behave to Hertz waves. It is inserted with a query, to indicate that its position in the table is not *certainly* known. It may possibly belong to some other column.

Electrical Theory of Vision.

And I want to suggest that quite possibly the sensitiveness of the eye is of the coherer kind. As I am not a physiologist I cannot be seriously blamed for making wild and hazardous speculations in that region. I therefore wish to guess that some part of the retina is an electrical organ, say like that of some fishes, maintaining an electromotive force which is prevented from stimulating the nerves solely by an intervening layer of badly conducting material, or of conducting powder with gaps in it; but that when light falls upon the retina these gaps become more or less conducting, and the nerves are stimulated. I do not feel clear which part is taken by the rods and cones, and which part by the pigment cells; I must not try to make the hypothesis too definite at present, though I hope it is obvious what I intend to suggest.

If I had to make a demonstration model of the eye on these lines, I should arrange a little battery to excite a frog's nerve-muscle preparation through a circuit completed all except a layer of filings or a single bad contact. Such an arrangement would respond to Hertz waves. Or, if I wanted actual light to act, instead of grosser waves, I would use a layer of selenium.

But the bad contact and the Hertz waves are the most instructive, because we do not at present really know what the selenium is doing, any more than what the retina is doing.

And observe that (to my surprise, 1 confess) the rough outline of a theory of vision thus suggested is in accordance with some of the principal views of the physiologist Hering. The sensation of light is due to the electrical stimulus; the sensation of black is due to the mechanical or tapping

DETECTORS OF RADIATION.

Physiological.	Chemical.	Thermal.	Electrical.	Mechanical.	Microphonic.
Eye.	Photographic Plate.	Thermopile.	Spark. (Hertz.)	Electrometer. (Blyth and Bjerknes.)	Selenium.(?)
×Frog's Leg (Hertz and Ritter.)	Explosive Gases.	Bolometer. (Rubens and Ritter.)	{ Telephone ; Air-gap and Arc. (Lodge.) }	Suspended Wires. (Hertz and Boys.)	Impulsion Cell. (Minchin.)
	Photoelectric Cell.	Expanding Wire. (Gregory.) Thermal Junction. (Klemencic.)	Vacuum Tube. (Dragoumis.) Galvanometer. (Fitzgerald.) Air-gap and Electroscope. (Boltzmann.) Trigger Tube. (Warburg and Zehnder.)		Filings. (Branly.) Coherer. Hughes and Lodge.

×The cross against the frog's leg indicates that it does not appear really to respond to radiation, unless stimulated in some secondary manner. The names against the other things are unimportant, but suggest the persons who applied the detector to electric radiation. The interrogation mark against Selenium indicates that its position in the microphonic column may be doubtful.

back stimulus. Darkness is physiologically not the mere cessation of light. Both are positive sensations, and both stimuli are necessary; for until the filings are tapped back vision is persistent. In the eye model the period of mechanical tremor should be, say, $\frac{1}{10}$th second, so as to give the right amount of persistence of impression.

No doubt in the eye the tapping back is done automatically by the tissues, so that it is always ready for a new impression, until fatigued. And by mounting an electric bell or other vibrator on the same board as a tube of filings, it is possible to arrange so that a feeble electric stimulus shall produce a feeble steady effect, a stronger stimulus a stronger effect, and so on; the tremor asserting its predominance, and bringing the spot back, whenever the electric stimulus ceases.

An electric bell thus close to the tube is, indeed, not the best vibrator; clockwork might do better, because the bell contains in itself a jerky current, which produces one effect, and a mechanical vibration, which produces an opposite effect, hence the spot of light can hardly keep still. By lessening the vibration—say, by detaching the bell from actual contact with the board, the electric jerks of the intermittent current drive the spot violently up the scale; mechanical tremor brings it down again. It must be clearly understood that electric jerks, due to the make and break of an ordinary current, are quite adequate to electrically stimulate a coherer in their neighbourhood. It is constantly to be noticed that a coherer responds best to excessively short sparks of a certain sharp quality.

You observe that the eye on this hypothesis is, in electro-meter language, heterostatic. The energy of vision is supplied by the organism; the light only pulls a trigger. Whereas the organ of hearing is idiostatic. I might draw further analogies between this arrangement and the eye, *e.g.*, about the effect of blows or disorder causing irregular conduction and stimulation, of the galvanometer in the one instrument, of the brain cells in the other.

A handy portable exciter of electric waves is one of the ordinary hand electric gas-lighters, containing a small revolving doubler—*i.e.*, an inductive or replenishing machine. A coherer can feel a gas-lighter across a lecture theatre.

Minchin often used them for stimulating his impulsion cells. I find that when held near they act a little even when no ordinary spark occurs, plainly because of the little incipient sparks at the brushes or tinfoil contacts inside. A Voss machine acts similarly, giving a small deflection while working up before it sparks : indeed, these small sparks are often more effective than bigger ones.

Demonstration of Ordinary Holtz Machine Sparks not exciting Tube : except by help of a polished knob.

And notice here that our model eye has a well-defined range of vision. It cannot see waves too long for it. The powerful disturbance caused by the violent flashes of a Holtz or Wimshurst or Voss machine it is blind to. The loud sparks have no effect on it. They are like infra-red radiation to the eye. If the knobs of the machine are well polished the coherer begins to respond again, evidently by reason of some high harmonics, due to vibrations in the terminal rods ; and these are the vibrations to which it responds when excited simply by an induction-coil. The coil should have knobs instead of points. Sparks from points or dirty knobs hardly excite the coherer at all. But hold a well-polished sphere or third knob between even the dirty knobs of a Voss machine, and the coherer responds at once to the surgings got up in that clean sphere.

Feeble short sparks again are often more powerful exciters than are strong long ones. I suppose because they are more sudden. This is instructively shown with an electrophorous lid. Spark it to a knuckle, and it does very little. Spark it to a clean knob held in the hand and it works well. But now spark it to an *insulated* sphere, there is some effect. Discharge the sphere, and take a second spark, without recharging the lid ; do this several times ; and at last, when the spark is inaudible, invisible, and otherwise imperceptible, the coherer some yards away responds more violently than ever, and the spot of light rushes from the scale.

If a coherer be attached by a side wire to the gas pipes, and an electrophorus spark be given to either the gas pipes or the water pipes or even to the hot-water system in any other room of the building, the coherer responds. It is surprising how far these impulses can be felt along an ordinary uninsulated wire or other conductor.

In fact, when thus connected to gas-pipes one day when I tried it, the spot of light could hardly keep still five seconds. Whether there was a distant thunderstorm, or whether it was only picking up telegraphic jerks, I do not know. The jerk of turning on or off an extra Swan lamp can affect it when sensitive. I hope to try for long-wave radiation from the sun, filtering out the ordinary well-known waves by a blackboard or other sufficiently opaque substance.

[I did not succeed in this, for a sensitive coherer in an outside shed unprotected by the thick walls of a substantial building cannot be kept quiet for long. I found its spot of

FIG. 19B.—A Portable Detector, B the Collecting Wire.

light liable to frequent weak and occasionally violent excursions, and I could not trace any of these to the influence of the sun. There were evidently too many terrestrial sources of disturbance in a city like Liverpool to make the experiment feasible. I don't know that it might not possibly be successful in some isolated country place; but clearly the arrangement must be highly sensitive in order to succeed.]

We can easily see the detector respond to a distant source of radiation now, viz., to a 5in. sphere placed in the library between secondary coil knobs; separated from the receiver, therefore, by several walls and some heavily gilded paper, as well as by 20 or 30 yards of space (Fig. 19.)

Also I exhibit (Fig. 19B) a small complete detector made by my assistant, Mr. Davies, which is quite portable and easily set

up. The essentials (battery, galvanometer, and coherer) are all
in a copper cylinder, A, three inches by two. A bit of wire, B,
a few inches long, pegged into it, helps it to collect waves. It
is just conceivable that at some distant date, say by dint of
inserting gold wires or powder in the retina, we may be
enabled to see waves which at present we are blind to.

Observe how simple the production and detection of Hertz
waves are now. An electrophorus or a frictional machine
serves to excite them ; a voltaic cell, a rough galvanometer,
and a bad contact serves to detect them. Indeed, they might
have been observed at the beginning of the century, before
galvanometers were known : a frog's leg or an iodide of
starch paper would do almost as well.

A bad contact was at one time regarded as a simple nuisance,
because of the singularly uncertain and capricious character of
the current transmitted by it. Hughes observed its sensitiveness to sound-waves, and it became the microphone. Now it
turns out to be sensitive to electric waves, if it be made of any
oxidisable medal (not of carbon),* and we have an instrument
which might be called a micro-something, but which, as it
appears to act by cohesion, I at present call a coherer. Perhaps
some of the capriciousness of an anathematised bad contact was
sometimes due to the fact that it was responding to stray electric radiation. (See Appendix III., pp. 109 and 111.)

The breaking down of cohesion by mechanical tremor is an
ancient process, observed on a large scale by engineers in railway axles and girders ; indeed, the cutting of small girders by
persistent blows of hammer and chisel reminded me the other
day of the tapping back of our cohering surfaces after they
have been exposed to the uniting effect of an electric jerk.

Receiver in Metallic Enclosure.

If a coherer is shut up in a complete metallic enclosure,
waves cannot get at it, but if wires are led from it to an outside ordinary galvanometer, it remains nearly as sensitive as

---

* FitzGerald tells me that he has succeeded with carbon also. My
experience is that the less oxidisable the metal, the more sensitive and also
the more troublesome is the detector. Mr. Robinson has now made me a
hydrogen vacuum tube of brass filings, which beats the coherer for
sensitiveness. July, 1894.

it was before (nearly, not quite), for the circuit picks up the waves and they run along the insulated wires into the closed box. To screen it effectively, it is necessary to enclose battery and galvanometer and every bit of wire connection; the only thing that may be left outside is the needle of the galvanometer. Accordingly, here we have a compact arrangement of battery and galvanometer coil and coherer, all shut up in a copper box (Fig. 19c). The galvanometer coil is fixed against the side of the box at such height that it can act conveniently on an outside suspended compass needle. The slow magnetic

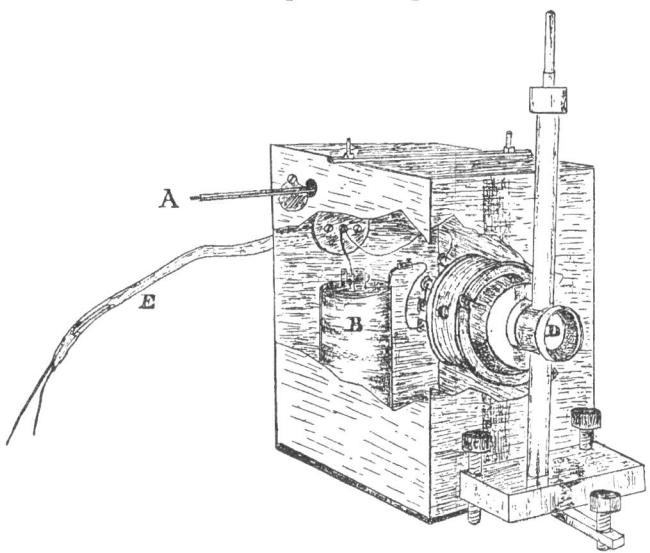

Fig. 19c.—Protected Detector. A is an occasional wire passing through shuttered aperture. E is a lead tube enclosing leading wires, as in Fig. 21.

action of the current in the coil has no difficulty in getting through copper, as everyone knows : only a perfect conductor could screen off that ; but the Hertz waves are effectively kept out by the sheet copper.

It must be said, however, that the box must be exceedingly well closed for the screening to be perfect. The very narrowest chink permits their entrance, and at one time I thought I should have to solder a lid on before they could be kept entirely out. Clamping a copper lid on to a flange in six places was not enough. But by the use of pads of tinfoil and tight clamping, chinks can be avoided, and the inside of the box becomes then electrically dark.

If even an inch of the circuit protrudes, it at once becomes slightly sensitive again ; and if a mere single wire protrudes through the box, not connected to anything at either end, provided it is insulated where it passes through, the waves will utilise it as a speaking tube, and run blithely in. And this happens whether the wire be connected to anything inside or not, though it acts more strongly when connected.

In careful experiments, where the galvanometer is protected in one copper box and the coherer in another, the wires con-

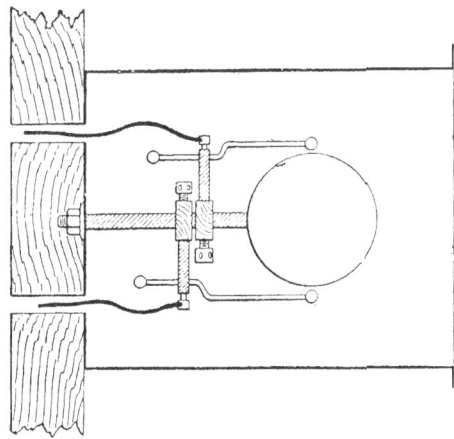

Fig. 20.—Spherical Radiator for emitting a Horizontal Beam, arranged inside a Copper Hat, fixed against the outside of a metal-lined Box, which contains induction coil and battery and key. One-eighth natural size. The wires pass into the box through glass tubes not shown.

necting the two must be encased in a metal tube (Figs. 19c and 21), and this tube must be well connected with the metal of both enclosures, if nothing is to get in but what is wanted.

Similarly when definite radiation is desired, it is well to put the radiator in a copper hat open in only one direction (Fig. 20), and in order to guard against reflected and collateral surgings running along the wires which pass outside to the exciting coil and battery, as they are liable to do, I am accustomed to put all the sending apparatus in a packing case lined with tinfoil, to the outside of which the sending hat (Fig. 20) is fixed, and to pull the key of the primary exciting circuit by a string from outside, so that not even key connections shall protrude, else exact optical experiments are impossible.

Even then, with the lid of the hat well clamped on, something gets out, but it is not enough to cause serious disturbance of qualitative results. The sender must evidently be thought of as emitting a momentary blaze of light which escapes through every chink. Or, indeed, since the waves are some inches long, the difficulty of keeping them out of an

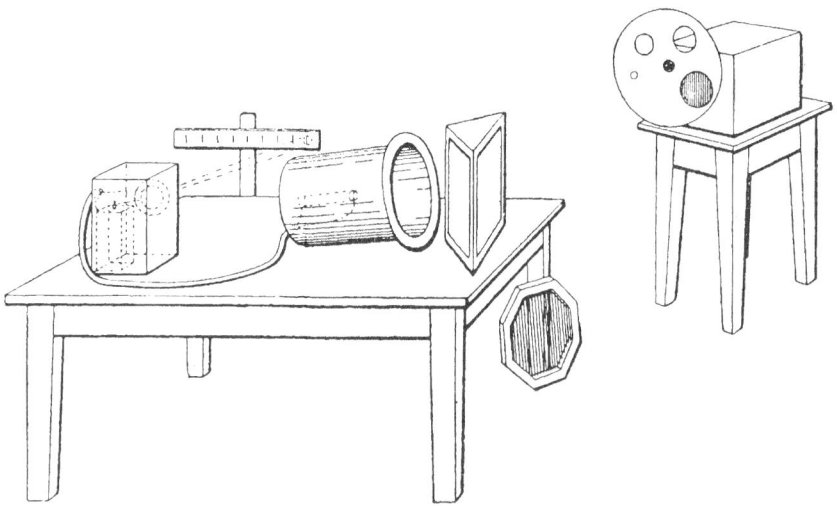

Fig. 21.—General arrangement of experiments with the Copper " Hat," showing Metal Box on a Stool, standing outside the Theatre. The Box is not exactly represented, but inside it the Radiators were fixed with a graduated series of apertures ; the Copper Hat containing the Coherer is seen on the Table with the Metal Box on the left of the Table containing Battery and Galvanometer Coil connected to it by a compo pipe conveying the wires, as in Fig. 19c ; the Lamp and Scale barely indicated at one side of the Table ; a Paraffin Prism ; and a Polarising Grid of copper wires stretched on a frame. [This figure is from a thumbnail sketch by Mr. A. P. Trotter, taken at the Lecture in 1894.]

enclosure may be likened to the difficulty of excluding sound ; though the difficulty is not quite so great as that, since a reasonable thickness of metal is really opaque. I fancied once or twice I detected a trace of transparency in such metal sheets as ordinary tinplate, but unnoticed chinks elsewhere may have deceived me. It is a thing easy to make sure of as soon as I have more time. [Tinplate is quite opaque   Lead paper lets a little through.]

One thing in this connection is noticeable, and that is how little radiation gets either in or out of a small *round* hole. A narrow long chink in the receiver box lets in a lot ; a round hole the size of a shilling lets in hardly any, unless indeed a bit of insulated wire protrudes through it like a collecting ear trumpet, as at A, Fig. 19c.

It may be asked how the waves get out of the metal tube of an electric gas-lighter. But they do not ; they get out through the handle, which being of ebonite is transparent. Wrap up the handle in tinfoil, and a gas-lighter is powerless.

OPTICAL EXPERIMENTS.

And now, in conclusion, I will show some of the ordinary optical experiments with Hertz waves, using as source either one of two devices : either a 5in. sphere with sparks to ends of a diameter (Fig. 19), an arrangement which emits 7in. waves but of so dead-beat a character that it is wise to enclose it in a copper hat to prolong them and send them out in the desired direction, or else a 2in. hollow cylinder with spark knobs at ends of an internal diameter (Fig. 12). This last emits 3in. waves of a very fairly persistent character, but with nothing like the intensity of one of the outside radiators.

As receiver there is no need to use anything sensitive, so I employ a glass tube full of coarse iron filings, put at the back of a copper hat with its mouth turned well askew to the source, which is put outside the door at a distance of some yards, so that only a little direct radiation can reach the tube. Sometimes the tube is put lengthways in the hat instead of crossways, which makes it less sensitive, and has also the advantage of doing away with the polarising, or rather analysing, power of a crossway tube.

The radiation from the sphere is still too strong, but it can be stopped down by a diaphragm plate with holes in it of varying size clamped on the sending box (right hand side of Fig. 21).

Reflection.

Having thus reduced the excursion of the spot of light to a foot or so, a metal plate is held as reflector, and at once the spot travels a couple of yards. A wet cloth reflects something, but a thin glass plate, if dry, reflects next to nothing, being, as is well known, too thin to give anything but "the black spot." I have fancied that it reflects something of the 3in. waves.

With reference to the reflecting power of different substances, it may be interesting to give the following numbers showing the motion of the spot of light when 8in. waves were reflected into the copper hat, the angle of incidence being about 45deg., by the following mirrors :—

Sheet of window glass	0 or at most 1 division.
Human body	7 divisions.
Drawing board	12 ,,
Towel soaked with tap-water	12 ,,
Tea-paper (lead ?)	40 ,,
Dutch metal paper...	70 ,,
Tinfoil	80 ,,
Sheet copper ...._.....................	100 and up against stops.

Refracting Prism and Lens.

A block of paraffin about a cubic foot in volume is cast into the shape of a prism with angles 75deg., 60deg., and 45deg. Using the large angle, the rays are refracted into the receiving hat (Fig. 21), and produce an effect much larger than when the prism is removed.

An ordinary 9in. glass lens is next placed near the source, and by means of the light of a taper it is focussed between source and receiver. The lens is seen to increase the effect by concentrating the electric radiation.

Arago Disc ; Grating ; and Zone-plate.

The lens helps us to set correctly an 18in. circular copper disc in position for showing the bright diffraction spot. Removing the disc, the effect is much the same as when it was present, in accordance with the theory of Poisson. Add the lens and the effect is greater. With a diffraction grating of copper strips 2in. broad and 2in. apart, I have not yet succeeded in getting good results. It is difficult to get sharp nodes and interference effects with these sensitive detectors in a room. I expect to do better when I can try out of doors, away from so many reflecting surfaces; indoors it is like trying delicate optical experiments in a small whitewashed chamber well supplied with looking-glasses; nor have I ever succeeded in getting clear concentration with this zone-plate having Newton's rings fixed to it in tinfoil. The coherer, at any rate in a room, does not seem well adapted to interference experiments; it is probably too sensitive, and responds even at the nodes, unless they are made more perfect than is easily practicable. But really there is nothing of much interest now in diffraction effects, except the demonstration of the

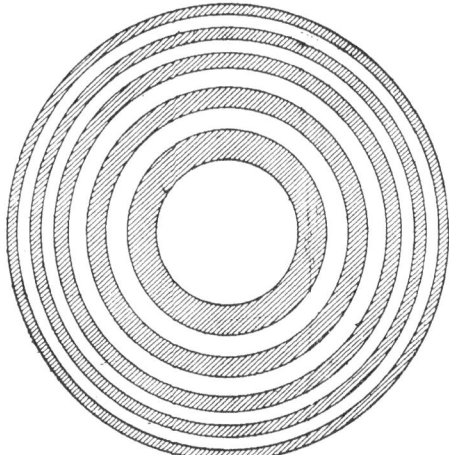

FIG. 22.—Zone-plate of Tinfoil on Glass. Every circular strip is of area equal to central space.

waves and the measure of their length. There was immense interest in Hertz's time, because then the wave character of the radiation had to be proved; but every possible kind of wave must give interference and diffraction effects, and their theory is, so to say, worked out. More interest attaches to polarisation, double refraction, and dispersion experiments.

Polarising and Analysing Grids.

Polarisation experiments are easy enough. Radiation from a sphere, or cylinder, or dumb-bell is already strongly

polarised, and the tube acts as a partial analyser, responding
much more vigorously when its length is parallel to the line
of sparks than when they are crossed ; but a convenient extra
polariser is a grid of wires something like what was used by
Hertz, only on a much smaller scale ; say an 18in. octagonal
frame of copper strip with a harp of parallel copper wires
(*see* Fig. 21, on floor). The spark-line of the radiator
(Fig. 20) being set at 45deg., a vertical grid placed over the
receiver reduces the reflection to about one-half, and a crossed
grid over the source reduces it to nearly nothing.

Rotating either grid a little rapidly increases the effect.
which becomes a maximum when they are parallel. The
interposition of a third grid, with its wires at 45deg., between
two crossed grids, restores some of the obliterated effect.

Radiation reflected from a grid is strongly polarised, of
course, in a plane normal to that of the radiation which gets
through it. They are thus analogous in their effect to Nicols,
or to a pile of plates.

The electric vibrations which get through these grids are at
right angles to the wires. Vibrations parallel to the wires
are reflected or absorbed.

*Reflecting Paraffin Surface ; Direction of Vibrations in Polarised
Light.*

To demonstrate that the so-called plane of polarisation of
the radiation transmitted by a grid is at right angles to the
electric vibration,* *i.e.*, that when light is reflected from the
boundary of a transparent substance at the polarising angle
the electric vibrations of the reflected beam are perpendicular
to the plane of reflection, I use the same paraffin prism as
before ; but this time I use its largest face as a reflector, and
set it at something near the polarising angle. When the line
of wires of the grid over the mouth of the emitter is parallel
to the plane of incidence, in which case the electric vibrations

---

* *Cf.* Trouton, in *Nature*, Vol. 39, p. 393 ; and many optical experiments
by Mr. Trouton, Vol. 40, p. 398. Since then the above described and
depicted apparatus for electro-optic experiments has been imitated in a
neat, compact form by Prof. J. Chunder Bose, of Calcutta, and with it he
has obtained many admirable and interesting optical results. See *Proc.*
Roy. Soc.

are perpendicular to the plane of incidence, plenty of radiation is reflected by the paraffin face. Turning the grid so that the electric vibrations are in the plane of incidence, we find that the paraffin surface set at the proper angle is able to reflect hardly anything. In other words, the vibrations contemplated by Fresnel are the electric vibrations; those dealt with by McCullagh are the magnetic ones.

Thus are some of the surmises of genius verified and made obvious to the wayfaring man.

END OF LECTURE.

NOTE WITH REFERENCE TO ELECTRIC WAVES ON WIRES.

It may be well to explain that in my Royal Institution lecture I made no reference to the transmission of waves along *wires*. I regard the transmission of waves in *free space* as the special discovery of Hertz ; though undoubtedly he got them on wires too. Their transmission along wires is, however, a much older thing. Von Bezold saw them in 1870, and I myself got quantitative evidence of nodes and loops in wires when working with Mr. Chattock in the session 1887-8 (*see*, for instance, contemporary reports of the Bath Meeting

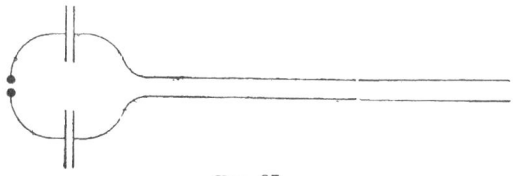

Fɪɢ. 23.

of the British Association, 1888, in *The Electrician*), and I exhibited them some time afterwards to the Physical Society, the wires themselves becoming momentarily luminous at every discharge except at the nodes, thus enabling the waves to be actually seen, having been made stationary by reflexion as in the corresponding acoustic experiment of Melde. This experiment does not appear to have been properly known (p. 78).

It may be worth mentioning that the arrangement frequently referred to in Germany by the name of Lecher (viz., that shown in Fig. 23), and on which a great number of experiments have been made, is nothing but a pair of Leyden jars with long wires leading from their outer coats, such as I constantly employed in these experiments. The wires from the outer coat in my experiment were very long, sometimes

going five or six times round a large hall, like telegraph
wires. And many measurements of wave length were thus
made by me at the same period as that in which Hertz was
working at Carlsruhe. The use of air dielectric instead of
glass permits the capacity to be adjusted, and also readily
enables the capacity to be small, and the frequency, therefore,
high ; but otherwise the arrangement is the same in principle
as had frequently been used by myself in the series of experiments called "the recoil kick" (*Proc.* Roy. Soc., June 1891,
Vol. 50, pp. 23–39). For these and other reasons no
reference has been made in my lecture to the work done
on wires by Sarasin and De la Rive ; nor to other excellent
work done by Lecher, Rubens, Arons, Paalzow, Ritter,
Blondlot, Curie, D. E. Jones, Yule, Barton, and other
experimenters.

APPLICATION OF THIS METHOD OF SIGNALLING AT A DISTANCE TO ACTUAL TELEGRAPHY.

---◆---

Although the method of signalling to a moderate distance through walls or other non-conducting obstructions by means of Hertz waves emitted from one station and detected by Branly filing tubes at another station was practised by the author and by several other persons in this country, it was not applied by them to actual telegraphy. The idea of replacing a galvanometer, which was preferably a well-damped or speaking galvanometer, by a relay working an ordinary sounder or Morse was an obvious one, but so far as the present author was concerned he did not realise that there would be any particular practical advantage in thus with difficulty telegraphing across space instead of with ease by the highly developed and simple telegraphic and telephonic methods rendered possible by the use of a connecting wire. In this non-perception of the practical uses of wireless telegraphy he undoubtedly erred. But others were not so blind, though equally busy ; and notably Dr. Alexander Muirhead foresaw the telegraphic importance of this method of signalling immediately after hearing the author's lecture on June 1st, 1894, and arranged a siphon recorder for the purpose. Captain Jackson also, at Devonport, made experiments for the Admiralty, and succeeded in telegraphing between ships in 1895 (or 1896). Prof. Popoff's telegraphic application in 1895 is mentioned on page 62.

By some chance a knowledge of the coherer method of detecting electric waves did not spread fast in Germany, the many German workers in Hertz waves continuing, for some time after 1894, the older and less efficient, though for metrical purposes often more convenient, mode of detecting them. But, in Italy, the work described in the preceding

lecture became well known, and the subject was developed
largely, especially by Prof. Righi, of Bologna, in the optical
direction. It was also developed in the same direction with
many most interesting results by Prof. Bose, of Calcutta, as
mentioned in the text. Prof. Righi made a large number of
experiments, which he has since described in an Italian
treatise, " Opticé Elettrica," and it appears that it was from
him that Signor Marconi learned about the subject, and
immediately conceived the idea of applying it to commercial
telegraphy. He appears to have worked at the subject
for a short time in Italy, aiming at getting the receiver
to be more satisfactory and dependable, and improving
the early form of Branly filings tube depicted on page 23
by greatly diminishing its size, bringing the terminals
closer together, and replacing the coarse borings by fine
filings. He also sealed them up in a vacuum, just as
the author did, as related on page 34. The only differences,
indeed, between his procedure and the author's during this
time were that Signor Marconi preferred nickel filings with a
little mercury and a low vacuum, whereas the author adhered
chiefly to iron and brass filings and to high vacua. At last
he brought it over to Dublin, where he was advised to take it
to the Chief of the Government Telegraphs, Mr. Preece, and
accordingly he took his, at that time, crude apparatus to the
Post Office in a sealed box. There was no point of novelty
in it at this stage.

With the powerful aid of the Post Office Signor Marconi
proceeded to develop his system of telegraphy on a large
scale ; and, sometimes failing, sometimes succeeding, gradually
increased the distance over which signalling was possible,
and especially began to develop from unpromising beginnings
his special method for long-distance, viz., the employment
of a sending and receiving conducting plate or other
small surface, at the top of a lofty pole, connected through
what was at that time supposed apparently to be the real
radiator, with the earth. This elevated plate, connected
as it now is through a simple spark gap with the earth, is an
obvious modification of a Hertz vibrator ; for it may be
regarded simply as a Hertz vibrator with its axis vertical, as
Hertz often used it, and with its lower plate replaced by the

earth, so as to double the available capacity ; but the action of
a pair of such elevated plates, connected through the earth
conductively and through the air inductively, as now used
by Marconi for sender and receiver respectively, is not
quite like that of a Hertz vibrator and a Hertz receiver
acting on one another by emitted radiation in the ordinary
way. If it were not the *same* earth to which the plates were
connected, they would have to act ordinarily by radiation,
but since it is the same earth, and that earth conducting
(possibly, indeed, with a submerged cable-sheath connecting
favourably-chosen stations), then the two elevated plates are
partially like the greatly separated terminals of a *single*
Hertz vibrator.

Only one of the plates is charged during a sending operation,
the other is at zero potential, but some trace of the electrostatic
lines from one plate may extend in curved lines to the other,
just as they extend to every elevated conductor within hail
of the sender in any direction.

Then comes the snap of the spark gap and the sudden
discharge, equivalent to the rush of an opposite charge of electricity suddenly into the sending plate, disturbing the electric
equilibrium at a distance—at any distance to which any trace
of electrostatic field had been able to reach—and giving a
kind of what is called in lightning a "*return stroke.*" The
effect of this on the distant plate and conductor must be
almost infinitesimal ; nevertheless, separating it from the
earth is the most sensitive detector to a minute sudden rush
or jerk of electricity that can be imagined, or that has hitherto
been invented,—the coherer. Accordingly, absurdly minute
though the disturbance is, the coherer feels it, instantly
increases in conductivity, works the relay, and gives the signal.
Every spark at the distant spark gap causes a similar rush
in or out of the distant elevated plate, and the receiving
plate collects such a fraction of this disturbance as to
stimulate the coherer and give a signal every time. Not
that it is to be supposed never to miss fire. At the present
time a coherer is not a rough instrument that can be left free
from expert attention with safety for a long time. There are
times when it goes on working for days or even weeks, but
there are other times when it gives trouble and needs some

form of attention.   Let us hope that these latter times will
become less frequent, and that the whole thing will become
quite dependable before long.   The pertinacious way in which
Mr. Marconi and his able co-operators have, at great expense,
gradually worked the method up from its early difficult and
capricious stage to its present great distances and comparative
dependableness is worthy of all praise.

Telegraphy by means of Hertz waves, though perhaps
chiefly developed in this country, has also been pursued
successfully by Prof. Slaby in Germany, who has attained
considerable distance over land, with its numerous obstacles,
and has published an account of his researches in a book
called " Funkentelegraphie "; while like success over land
has been attained by M. E. Ducretet, M. Blondel and others
in France.   M. Ducretet has, indeed, put on the market a
compact apparatus whereby beginners can readily try their
hands at this mode of signalling; as well as a large-scale
apparatus like that employed by Lieutenant Tissot for light-
house signalling on the coast of Brittany.

The filings tube now chiefly employed by the author is of
the following form :—It is a sealed glass tube containing
carefully selected iron filings, and exhausted to the highest
vacuum.   Close together are two little silver globes melted
each on its own platinum wire terminal, which are connected
with convenient screws on an ebonite stand.   The filings are
adjusted so as just to cover the two silver globes, and no
more ; a pocket, or reservoir, however, is sometimes provided
whereby more or fewer filings can be easily introduced into
the working compartment for experimental purposes.   This
pocket serves to fix the whole tube to its ebonite body, which
is provided with a clamp to attach it to the stiff spring, or
movable lever, or other form of support, through which it
is to receive the mechanical shocks necessary to restore or
decohere it after an electrical stimulus.

The usual plan is to employ an electrical hammer to rap
strongly on a stiff brass spring to which the ebonite is
clamped, but another plan is to attach the coherer to a lever
tilted strongly by an electromagnet after the fashion of a
sounder.   A rapid succession of gentle taps is often better
than one violent one, but experience is the best test of the

kind of restoration wanted, for it depends a good deal on the strength of the electrical stimulus. There are methods of dispensing with this decohering operation altogether.

After a fairly strong electric stimulus all the filings are stuck together into a sort of mat, and nothing but a thorough shaking up will pull them asunder again. A still more violent electric shock may indeed have a decohering effect, but it is not a plan to be recommended, for it seems to be a heat effect, akin to the blowing of a fuse.

For protecting a coherer from undesired stimuli, *e.g.*, from the radiator at its own station, the general method is described on page 35, &c., and the details of it, with the necessary switch for changing over from sending to receiving, are mentioned further on (page 60). But by referring to page 106 it will be seen that M. Branly had already employed such a protecting case, and had worked details out admirably.

Recently Signor Tommasina has shown that, if one end of a short rod or wire be dipped into filings while sparks are occurring in the neighbourhood, the filings adhere to it and to each other, and with care a long string of them can be picked up. The author has examined the behaviour of filings under electrical influence on a glass plate in a microscope, and their movements towards the formation of a complete conducting bridge between the tinfoil terminals together with their disjunctive behaviour when the electrical stimulus is too strong, the thorough cohesion set up by a succession of electrical stimuli, and the partial or complete disruption by an appropriate mechanical stimulus is instructive.

An earlier and most important telegraphic application, based upon information given in the preceding lecture, was made in 1895 by Prof. Popoff, of Russia, and will be mentioned shortly (*see* page 62). I now proceed to developments of syntonic or attuned telegraphy on the true Hertz-wave principle, the preliminary experiments on which are mentioned above in connection with the figures on page 27.

FURTHER DEVELOPMENTS IN THE TELEGRAPHIC DIRECTION.

———

SYNTONIC TELEGRAPHY.

In the present state of the law in this country it appears to be necessary for a scientific man whose investigations may have any practical bearing to refrain from communicating his work to any scientific society, or publishing it in any journal until he has registered it and paid a fee to the Government under the so-called Patent Law. This unfortunate system is well calculated to prevent scientific men in general from giving any attention to practical applications, and to deter them from an attempt to make their researches useful to the community. If a scientific worker publishes in the natural way, no one has any rights in the thing published ; it is given away and lies useless, for no one will care to expend capital upon a thing over which he has no effective control. In this case practical developments generally wait until some outsider steps in and either patents some slight addition or modification, or else, as sometimes happens, patents the whole thing, with some slight addition. If a scientific worker refrains from publishing and himself takes out a patent, there are innumerable troubles and possible litigation ahead of him, at least if the thing turns out at all remunerative ; but the probability is that, in his otherwise occupied hands, it will not so turn out until the period of his patent right has expired.

Pending a much-to-be-desired emendation of the law, whereby the courts can take cognisance of discoveries or fundamental steps in an invention communicated to and officially dated by a responsible scientific society, and can thereafter award to the discoverer such due and moderate recompense as shall seem appropriate when a great industry has risen on the basis of that same discovery or funda-

mental invention—pending this much-to-be-desired modification of the law, it appears to be necessary to go through
the inappropriate and repulsive form of registering a claim
to an attempt at a monopoly. The instinct of the scientific
worker is to publish everything, to hope that any useful
aspect of it may be as quickly as possible utilised, and to
trust to the instinct for fair play that he shall not be the
loser when the thing becomes commercially profitable. To
grant him a monopoly is to grant him a more than doubtful
boon ; to grant him the privilege of fighting for his monopoly

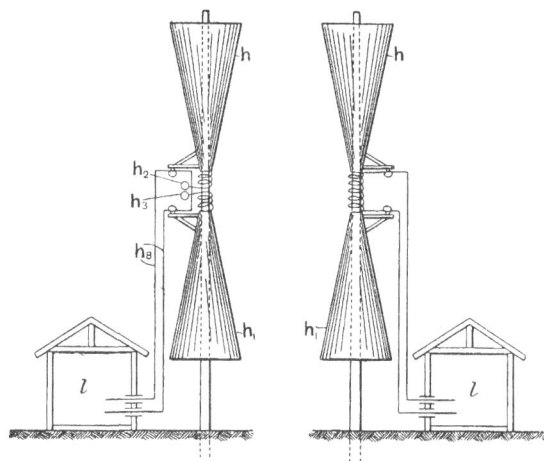

Fig. 24 (Fig. 5 of Specification 11,575/97).—Syntonic Radiator, adapted for
sending and for receiving.

is to grant him a pernicious privilege, which will sap his energy,
waste his time, and destroy his power of future production.

However, the author, in consultation with friends, decided
that registration was, under present conditions, necessary,
and, accordingly, for his attempt at syntony and other
improvements in the Hertz wave method of signalling, he
can refer here to certain patents taken out, in conjunction
chiefly with Dr. Alexander Muirhead, his co-worker, which are
numbered respectively as follows :—

(1) 11,575 of 1897, wherein is described the general
syntonic principle and the mode of prolonging the duration

of the vibrations emitted by a radiator or by a receiver. This
is done by adding to it electromagnetic inertia (that is, a self-
induction coil) in such a way as to lessen its radiating power,
converting its type of emission from something like a whip-
crack into something more like that of a struck string. (Not
pushing it so far as to make it like a *fork*, though that
could be done if desired : see *Journal* Inst.E.E., December,
1898.)   But too prolonged a duration of vibration is not
desirable, for it can only be obtained at the expense of
radiating power.   For the most distant signalling the single
pulse or whip-crack is the best, and this is what in practice

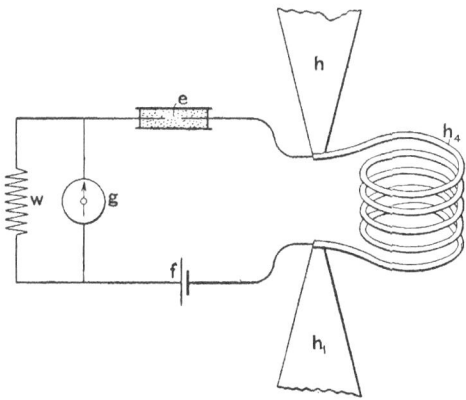

Fig. 25 (Fig. 13 of Specification 11,575/97). —Diagram of connections for
Syntonic Receiver ; *e* being coherer and *w* a non-inductive conducting or
capacity shunt, to eliminate the self-induction of the receiving instrument.

has hitherto always been employed ; but, with it, tuning is of
course impossible.   A radiator with several swings is less
violent at its first impulse than is a momentary emitter ;
but then the lessened emitting power of a radiator is to
be compensated by a correspondingly prolonged duration of
vibration on the part of the receiver or absorber, thus
rendering the radiator susceptible of tuning to a special
similarly-tuned receiver or resonator.   The tuned resonator
is then to respond, not to the first impulse of the radiator, but
to a rapidly worked up succession of properly timed impulses ;
so that at length, after an accumulation of two or three, or

perhaps four, swings, the electrostatic charges in its terminal plates become sufficient to overflow and spit off into the coherer, thereby effecting its stimulation and giving the signal. A resonator not properly tuned—*i.e.*, one tuned to some different frequency of vibration—would not be able to

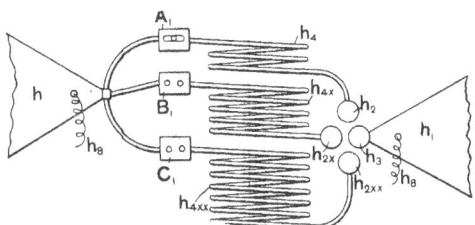

Fig. 26 (Fig. 10 of Specification 11,575/97).—Interchangeable Self-Induction Coils for signalling to different stations.

accumulate impulses, and hence would not respond, unless of course it were so much too near the radiator that the very first swing stimulated it sufficiently to disturb the coherer ; in which case, again, there is no room for tuning. The two points to attend to for syntonic discrimination are : (*a*) that

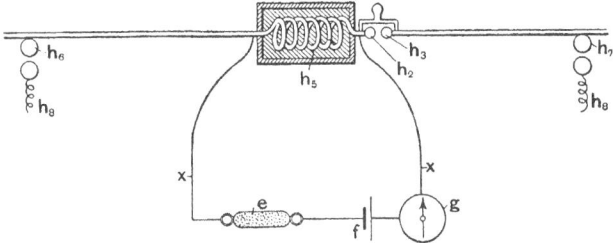

Fig. 27 (Fig. 3 of Specification 11,575/97).—Diagrammatic representation of Syntonic Radiator and Receiver. The middle spark gap h_2 h_3 is unnecessary, though sometimes helpful. The main charging is done by impulsive rush at the outside knobs.

the receiver shall not be so near the emitter as to feel its impulses too easily, *i.e.*, without accumulation ; (*b*) that the properly-tuned receiver shall be so arranged that it can work up and accumulate the impulses of the radiator, and before attaining its maximum swing can overflow into the coherer associated with it and thus give the signal.

The general appearance of a pair of signalling stations on this plan is shown in Fig. 24, where the huts contain the sending and receiving instruments. The self-induction coil joining the two capacity-areas is better depicted in Fig. 25, which also shows one mode of joining up the coherer to a syntonic receiver. (The galvanometer and shunt are, of course, merely typical of any kind of telegraphic instrument whatever.)   Fig. 26 indicates one form of sender with three alternative syntonising coils for speaking to three

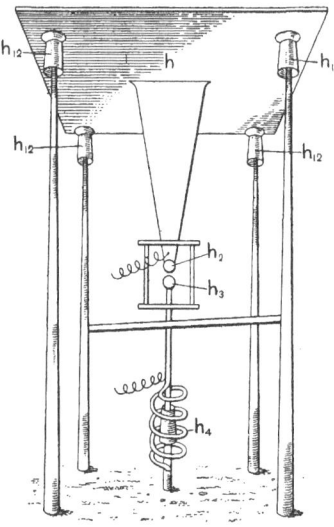

Fig. 28 (Fig. 7 of Specification 11,575/97).—Syntonic Radiator with earth connection arranged for sending.

distant attuned stations.   Fig. 27 shows a radiator arranged for receiving, but illustrates another method of charging, and one frequently employed by the author, viz., the method by impulsive rush (compare Figs. 11, 12 and 19, on pp. 14 and 25 of this book).   The terminals of the Ruhmkorff coil are here connected, not to the capacity areas direct, but to a pair of knobs near the centre of gravity of each area, so that when the discharge occurs each area is suddenly charged oppositely, and the two opposite charges are left to surge into one another and set up the oscillations.   This impulsive method of charging is essentially that adopted in the spherical whip-crack emitter depicted in Fig. 19 (p. 25, *ante*), the two poles of the sphere having but small capacity and being joined by as thick a conductor as the equator of the sphere. But for such a radiator as is indicated in Fig. 24 or Fig. 27 the author commonly found that a third short spark gap in the middle was an improvement, and so, as is well known, did Prof. Righi find it, and embodied it in his well-known double-sphere double-knob emitter.

The specification also contains figures of earth-connected forms of radiators, with or without self-induction coils, of

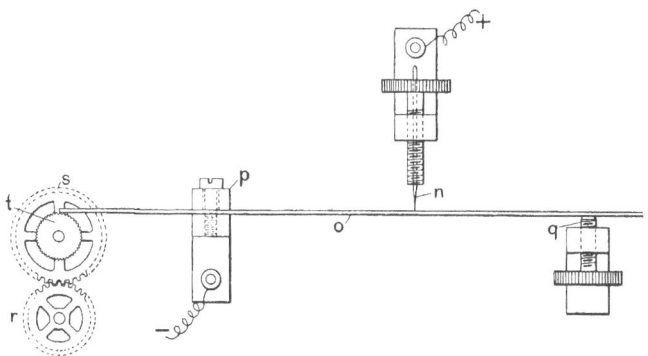

Fig. 29 (Fig. 12 of Specification 11,575/97).—Single-point Coherer, with clockwork Tapper-back operating on the projecting end of the spring clamped at P and lightly touching a needle-point *n*.

which Fig. 28 may be here reproduced; and likewise a modification of the point coherer depicted in Fig. 17, on page 22 (*see* Fig. 29, and also fig. on page 27), where the spiral wire spring is replaced by a piece of straight watch spring, clamped at one end, adjusted by a screw at the other, and lightly touched by a needle point at its middle; a very gentle tapping-back stimulus being provided in the form of a clockwork or other mechanically-driven motor grazing lightly against one end of the spring protruding beyond the clamp for the purpose.

Fig. 30 shows a coherer inserted in a secondary or transformer circuit, and operated inductively by the oscillations of

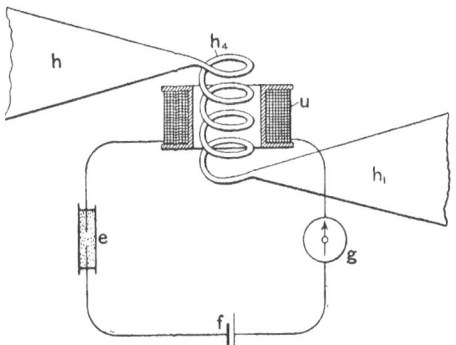

FIG. 30 (Fig. 14 of Specification 11,575/97).—Another diagram of connections for Syntonic Receiver, with Coherer in a secondary or transformer circuit ; a conducting or a capacity shunt for the telegraphic instrument being applicable as before.

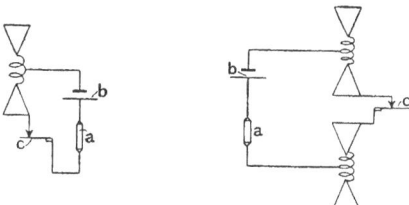

FIGS. 31 and 32 (Figs. 5 and 6 of Specification 18,644/97).—Modes of connecting a Coherer to one or to a pair of Syntonic Radiators so that it may feel their *electrostatic* disturbance.

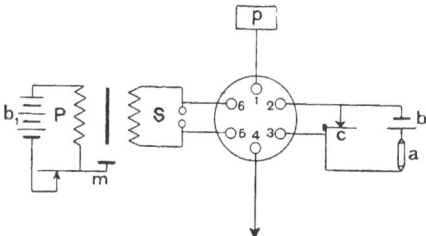

FIG. 33 (Fig. 11 of Specification 18,644/97).—Actual connections for a Sending and Receiving Station on the plan shown in Fig. 37. Left-hand side shows spark sending, right-hand side shows Coherer receiving.

the receiver, which are thus transformed up and raised in potential.

(2) No. 16,405, 1897, wherein are described chiefly various practical methods of decohering, by means of cams and otherwise, which are appropriate when working rapidly with automatic transmitter and siphon recorder.

(3) No. 18,644, 1897, represents different ways of connecting up a coherer to a syntonic resonator, so as to get the benefit of its overflow without interfering with the working up of the electric oscillations, *e.g.*, Figs. 31, 32 and 33. It also shows a plan for constantly decohering by a rapidly revolving cam a number of coherers in parallel, so that one at least is always

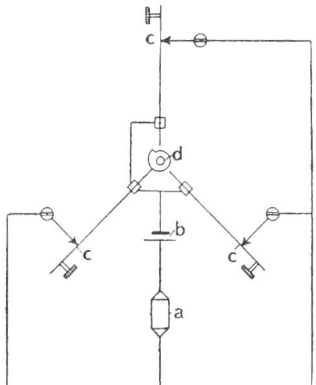

Fig. 34 (Fig. 1 of Specification 18,644/97).—Single-point Coherers in parallel, with successive decoherence.

ready to receive an impulse (Fig. 34). Further, it arranges to utilise the earth or a cable sheath, or other uninsulated conductor, for the purpose of conveying electric impulses to a distance (Figs. 35, 36, 37 and 38). And next it is arranged to assist the coherer to feel the full effect of any electric jerk by shunting out the battery and galvanometer, which are necessarily in series with it, by means of a condenser of moderate capacity (Fig. 35), which also shows a self-induction mode of sending a stimulus along an uninsulated line. This condenser obstructs all steady currents, such as give the signal, but it transmits freely any momentary electric impulses, such as stimulate a coherer.

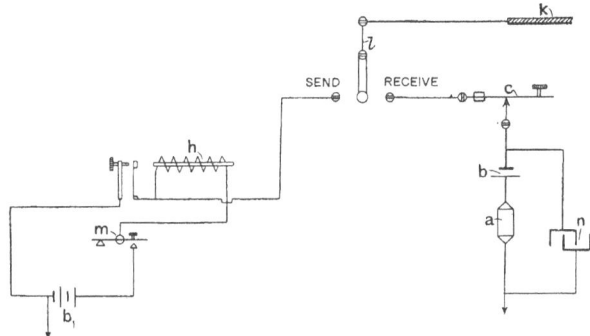

FIG. 35 (Fig. 3 of Specification 18,644/97).—A self-induction method of sending jerks into a badly insulated line, and arrangement for detecting such jerks by a single-point Coherer.

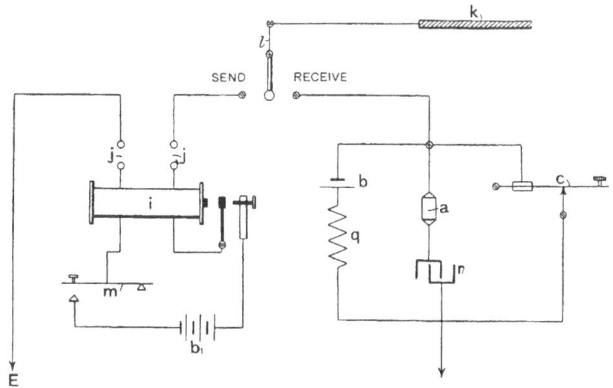

FIG. 36 (Fig. 4 of Specification 18,644/97).—Another arrangement for sending jerks into a bare or badly insulated line, and connections for Coherer detection.

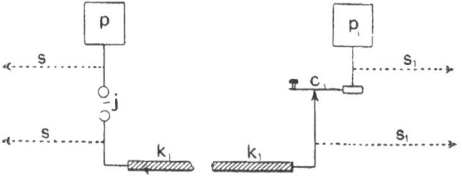

FIG. 37 (Fig. 10 of Specification 18,644/97).—Another mode of sending a jerk from a spark-gap at j into a badly insulated cable or other conductor, which is connected at the other end to a Coherer, the circuit being completed inductively through the air by means of the areas p, p_1. The dotted lines s represent the switch connection of Fig. 33.

(4) No. 29,069, 1897. In this patent various methods of connecting up the shunting condenser, whose object it is to transmit all jerks undiluted to the coherer, are shown, all adapted to work with a syntonic resonator (Fig. 39). There is also shown a complete switch (Fig. 40) for effecting the

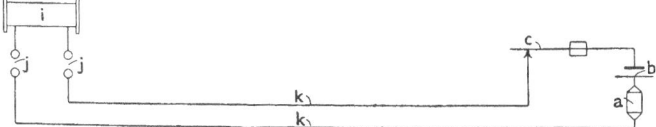

FIG. 38 (Fig. 13 of Specification 18,644/97).—Another method of signalling through a pair of imperfect conductors, such as gas and water pipes *i*, without the above elevated inductive connection.

transition from "sending" to "receiving," exposing the coherer to the full effect of the distant radiator, and completely protecting and isolating it from its home radiator; the switch being so arranged that signalling is impossible unless the home coherer is protected. A rotating commutator

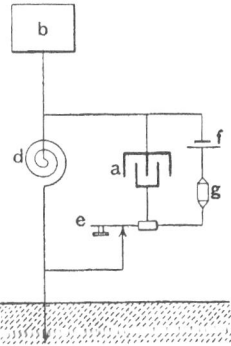

FIG. 39 (Fig. 3 of Specification 29,069/97).—Diagram of Coherer connection to Syntonic Collector, with capacity shunt for telegraphic instrument.

is also shown, whose object is to expose the coherer to the full influence of a receiver, especially of a non-syntonic receiver or simple collector, without its being shunted or otherwise interfered with by the telegraphic apparatus; to which, however, immediately afterwards the rotating commutator connects it, and then effects the tapping-back.

Connections are shown (Fig. 41) for a complete sending and receiving station on this plan with a syntonic radiator and resonator indicated (though not to scale). But with syntonic resonators the revolving commutator method is not found to be necessary ; the sending and receiving switch, together with the closed box for protecting the coherer in an instantly accessible manner is therefore the chief feature of this diagram.

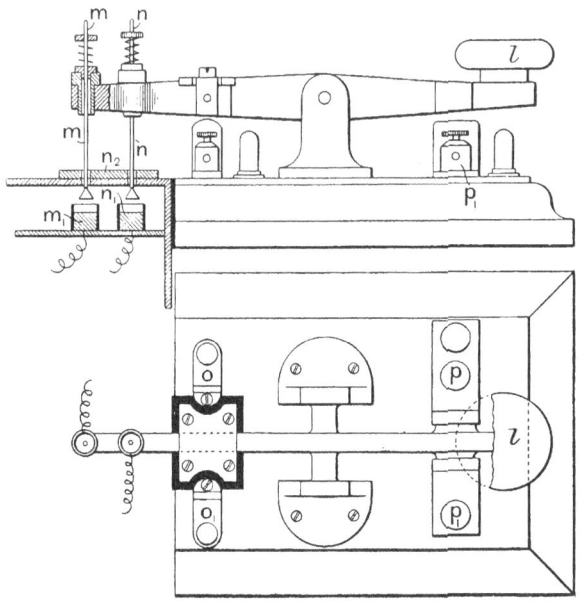

Fig. 40 (Fig. 6 of Specification 29,069/97).—Switch at a Sending and Receiving Station, to change all the connections with a protected Coherer from receiving to sending by depressing the knob *l.*

EARLIER TELEGRAPHIC ADVANCES.

In April, 1895, a communication was made to the Russian Physical Society by Prof. A. Popoff, of the Torpedo School, Cronstadt, Russia, and appears in the *Journal* of that Society for January, 1896. In this communication the use of an elevated wire and of a tapper-back worked through a relay by the coherer current are clearly described, and signalling was effected for a distance of 5 kilometres ($3\frac{1}{2}$ miles).

An extract from this communication is given in *The Electrician* for December, 1897, Vol. XL., page 235, and from it we reproduce Fig. 42, illustrating the tapping back arrangement.

The following extracts from this paper may also be quoted :—

"On using a sensitive relay in the circuit with the coherer tube, and an ordinary electric bell in the other circuit of the

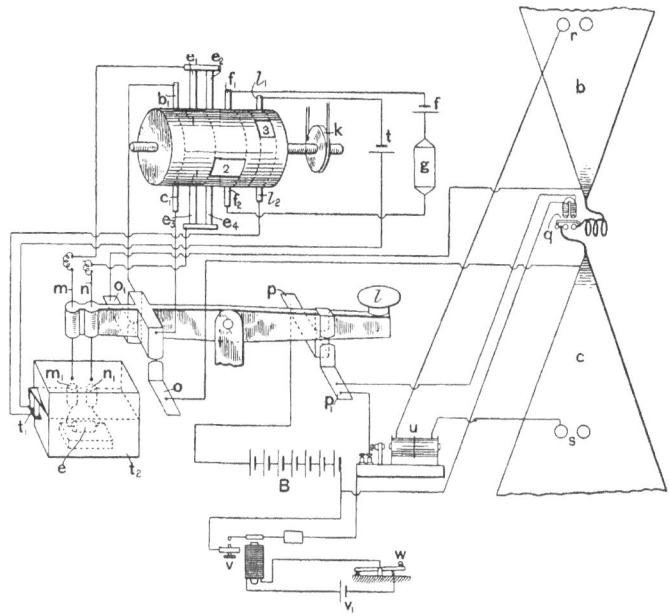

Fig. 41 (Fig. 7 of Specification 29,069/97) —Diagram of connections at a protected Coherer Station with Syntonic Radiator and Collector.

relay, for sound signals and as an automatic tapper for the coherer, I obtain an apparatus which exactly answers every electric wave by a short ring, and by rhythmical strokes if electric vibrations be excited continuously.

"On connecting an electro-magnetic recorder in parallel with the bell, tracing a straight line along the paper band which is moved by a 12-hour clockwork cylinder, I obtain an instrument registering by a cross line on the moving band

every electric wave that reaches the coherer from across the
atmosphere.   Such an apparatus was placed at the Meteorological Observatory at St. Petersburg in July, 1895, one of
the electrodes of the coherer being connected by an insulated
wire with an ordinary lightning conductor, the other electrode
of the tube-coherer being connected with the ground."

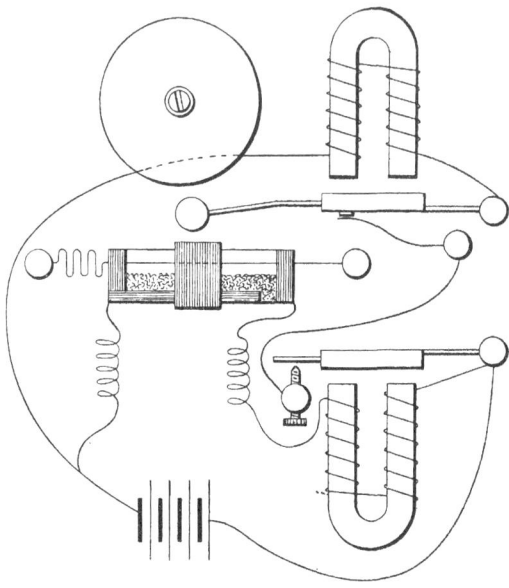

F<small>IG</small>. 42 (Fig. 2 on p. 235 of *The Electrician*, Vol. XL.).—Method of
automatic tapping back by relay-current employed for telegraphy by
Prof. Popoff in 1895.

Prof. Popoff then goes on to say that his apparatus works
well as a lightning recorder, and that he hopes it can be used
for signalling to great distances.   He says :—
    " I can detect waves at the distance of one kilometre if I
employ as sender a Hertz vibrator with 30 centimetre spheres,
and if I use the ordinary Siemens relay ; but with a Bjerknes
vibrator 90 centimetres diameter, and a more sensitive relay,
I reach five kilometres of good working."
    Thus it is plain that Prof. Popoff employed the elevated
wire as receiver in 1895, but did not employ it as sender.

In 1897 Prof. Slaby, of Berlin, published (in German) a
book called "Spark Telegraphy," in which he described his
success in signalling from 3 to 13 miles across land. From
this book we take the following illustrations of the coherer
and its connections :—

Fig. 43 shows the coherer tied on to a glass tube, by which it
is supported.

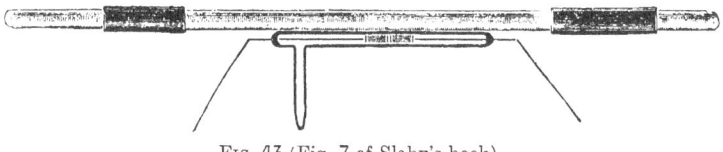

FIG. 43 (Fig. 7 of Slaby's book).

Fig. 44 shows the simplest form of its connection to a one-
cell battery A and a polarised relay B, which switches on
another battery of several cells *a* operating the Morse instrument or electric bell or sounder *b* and also the tapper-back *c*,
the hammer of which raps gently on the coherer tube at every
signal.

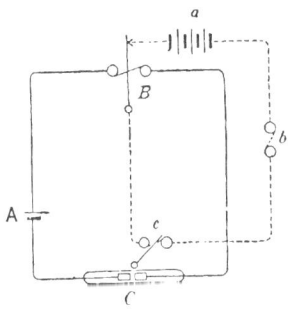

FIG. 44 (Fig. 8 of Slaby's book).—Slaby's arrangement of Coherer and of
tapper-back and relay connections.

The actual apparatus is depicted in two views, Figs. 45 and
46, where will be recognised on the left-hand side the coherer
and tapper-back ; in the middle the batteries, both for relay
and for coherer circuits ; and on the right-hand side a relay
and the signalling or calling instrument, in this case shown
as an ordinary electric bell.

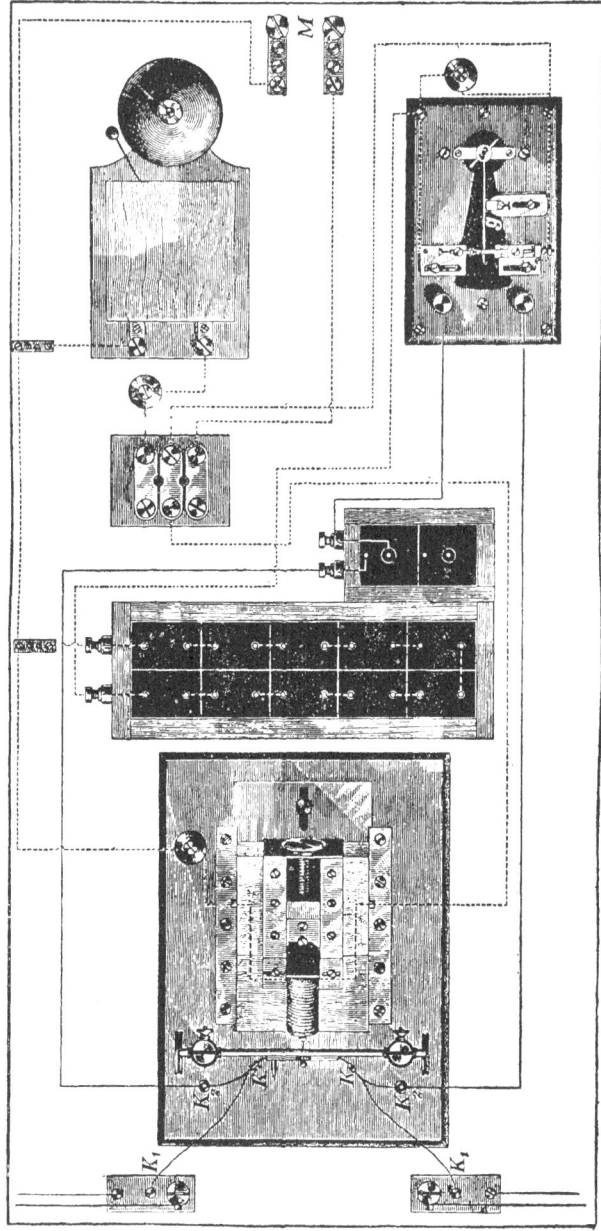

Fig. 45 (Fig. 16 of Slaby's book).—View of Slaby's Receiving Apparatus, with call-bell rung by relay, or with Morse instrument joined on to terminals M, and switch to change from Calling to Signalling. K K are the terminals of elevated wire and earth, and the Coherer and Tapper-back are close to them.

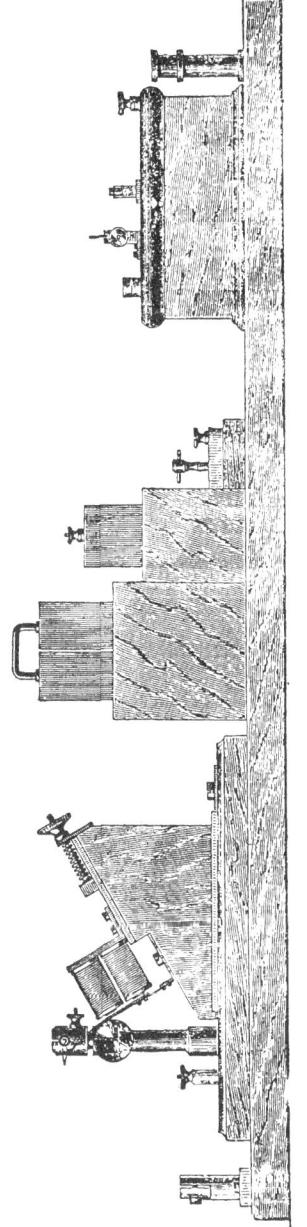

FIG. 46 (Fig. 17 of Prof. Slaby's book).—An elevation view of Prof. Slaby's same Apparatus, showing the electromagnet and hammer of the tapper-back worked by relay current from local battery, as in Popoff's plan of 1895.

A Morse instrument is to be connected to the terminals M, and either it or the bell can be switched into the circuit at pleasure. The form of relay depicted is special to Slaby, but the rest of the arrangements are practically identical with those shown by Marconi at Dover.

Fig. 47 gives a diagram of the actual connections.

Fig. 48 is a picture of one of Slaby's signalling stations, showing the way the elevated wire enters the building.

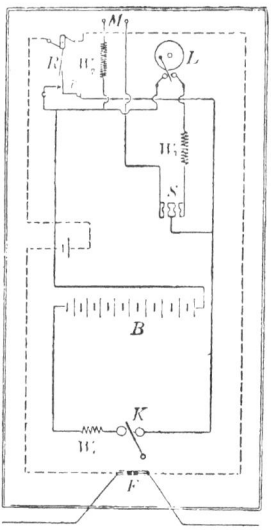

Fig. 47 (Fig. 19 of "Spark Telegraphy").—Diagram of Slaby's connections in the above apparatus. F is the coherer and K the tapper-back.

During September, 1899, the Marconi method of signalling to long distances was demonstrated before the British Association at Dover. The chief feature of the installation was the elevated wire supported by a mast, and terminating at the top in a small conductor, which is usually made of wire netting, and is suspended from an insulating rod. The lower end of this elevated wire passed into the building through an aperture, and was connected to one terminal of the usual Ruhmkorff coil, the other terminal of which was earthed. The signalling key was of the simplest description, being nothing more than a well-insulated Morse key worked by hand and causing a make-and-break in the primary circuit of the coil. The ordinary trembling break of the induction coil was at work in the usual way, so that while the signalling key was depressed continuously there was a torrent of sparks between the knobs of the secondary. This method of signalling was identical with that employed by everyone since the time of Hertz, except that, instead of connecting the secondary terminals to two insulated plates, one was now connected to earth and the other to a small insulated conductor at considerable elevation.

Fig. 48 (Fig. 11 of Prof. Slaby's book on " Spark Telegraphy ").

From this mast in the town of Dover (Fig. 49) signals could be sent to another loftier mast at the South Foreland (Fig. 50), where it is itself elevated by chalk cliffs far above the sea. From this South Foreland station, which was similar in all essential respects to the Dover station, except that its elevation was greater, messages could be sent and received to and from a station near Boulogne, on the coast of France, and to and from the East Goodwins lightship. The signalling was slow, but appeared dependable, and the simplicity of all the arrangements was remarkable (Fig. 51).

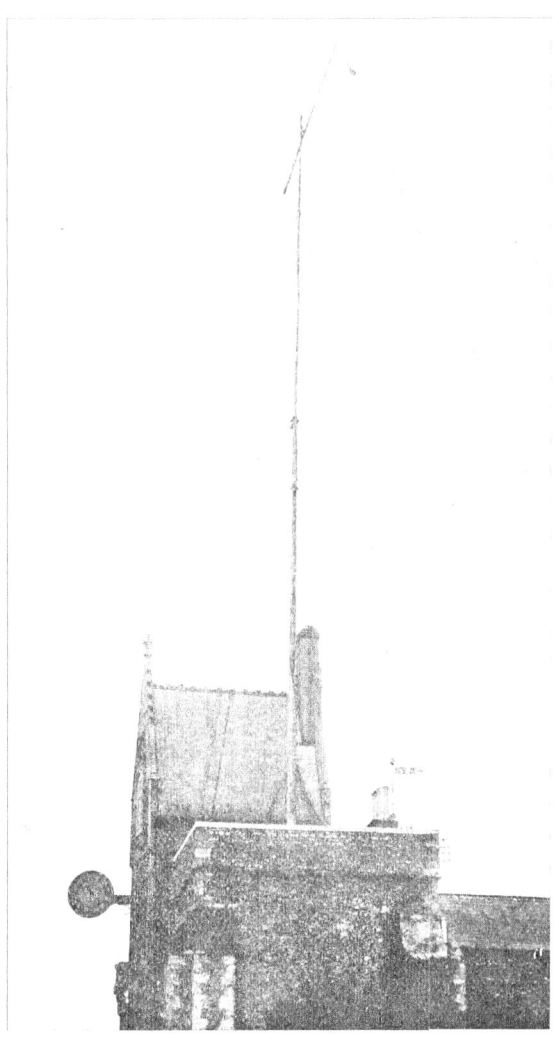

FIG. 49 (p. 762, *The Electrician,* Vol. XLIII.).—Marconi Signalling Mast
at Dover Town Hall.

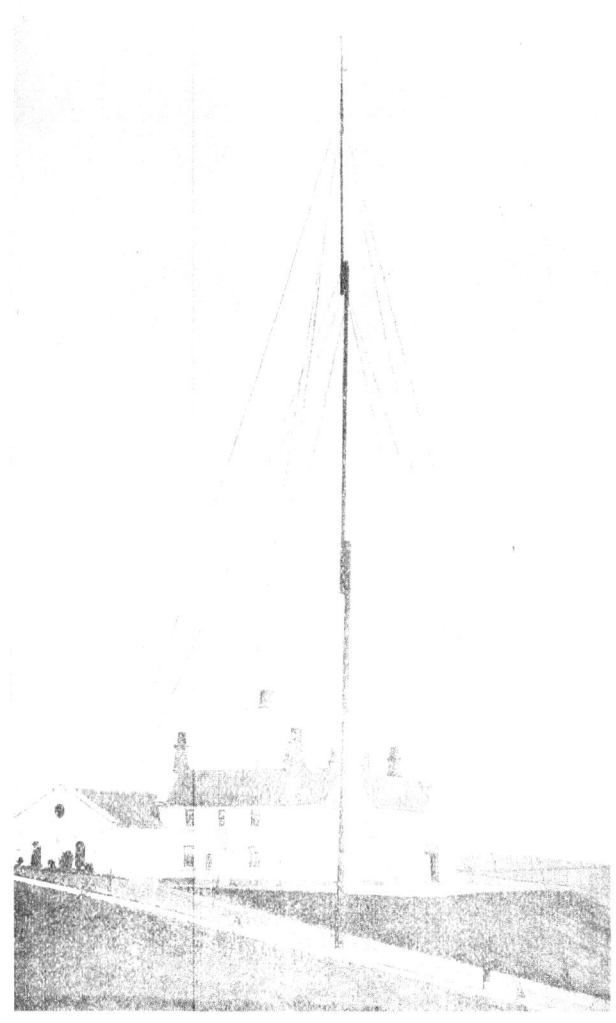

Fig. 50 (Fig. 1, p. 7, *The Electrician*, Vol. XLIII.).—Mast at South Foreland, from which Signals went to a similar Mast at Wimereux, near Boulogne.

Concerning the receiving apparatus there is little to be said, since it is in essence the same as that which has already been described. It consists of a coherer of the plug tube pattern, something like that depicted on page 23, but excessively reduced in size, the glass tube being the size of a quill, the two silver plugs close together separated only by a very few nickel filings. This tube is mounted so that it can be struck after each signal by a light electric hammer worked by a current from a local battery switched on by a Siemens' polarised relay, which is itself actuated by the coherer current. Whenever the coherer receives a signal the same current that works the tapper works also the Morse instrument standing on the table alongside, and records a short or a long signal on the tape. The coherer with its tapper, the polarised relay, and the battery (a few dry cells) are all enclosed in one oblong iron box, through an aperture in which the lower end of the elevated wire can be inserted and brought into direct connection with the coherer.

To change from transmitting to receiving nothing is needed but the detachment of this wire from the Ruhmkorff coil terminal and its insertion through the aperture of the enclosing box so as to touch the coherer circuit. The object of the box is, of course, the protection of the coherer from undesired disturbances, exactly as described on page 34, and the collecting wire has the function there described likewise.

The electric tapper-back is also mentioned on page 31, but not as being operated through a relay by the coherer circuit's own current. This last improvement seems to have been devised and employed by Prof. Popoff at Cronstadt in 1895 (*see* Fig. 42). No doubt it was arrived at independently again by Mr. Marconi and the telegraph officials who assisted him in his early experiments in this country.

The other box shown in Fig. 51 is probably a stand-by in case of accident.

It is difficult to imagine a simpler contrivance, and it appeared to work at Dover dependably, the messages coming out slowly in ordinary dots and dashes, the torrent of sparks being sufficiently rapid not to necessitate the breaking up of the dash into a series of dots. The sluggishness of the Morse

FIG. 51 (p. 761, *The Electrician*, Vol. XLIII.).—Apparatus for Sending and Receiving, shown by Prof. Fleming to the British Association at Dover.

instrument or the relay, or the circuit as a whole, enabled this excellent result to be attained with apparent ease.

A diagram of Marconi's connection of sensitive tube to the relay and tapper-back and Morse instrument, where W represents the elevated wire, is given in Fig. 52.

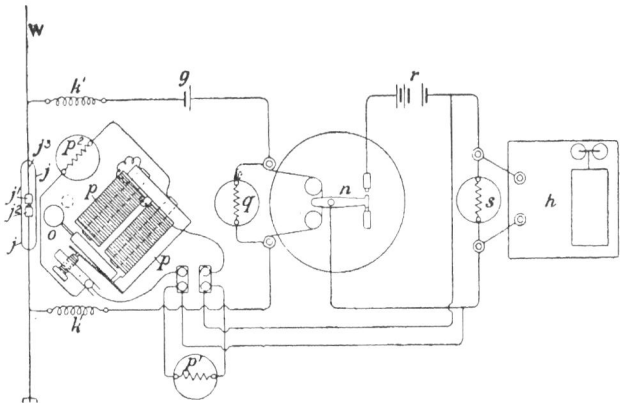

FIG. 52 (Fig. 2, of p. 691, *The Electrician*, Vol. XLII.).—Diagram of the connection of Relay and Tapper-back and Morse Instrument, as given in Mr. Marconi's Paper in the *Journal* of the Inst. Elec. Engineers for April, 1899 ; the relay being an ordinary Siemens polarised relay.

The mast at the South Foreland was stated to be 150ft. high, but the cliff on which it stands must be at a still greater elevation above the sea. It was from this station that the real distant signalling was performed, and probably not from the lower mast at Dover.

THE HISTORY OF THE COHERER PRINCIPLE.

*The following, written by Dr. Oliver Lodge, appeared in "*THE ELECTRICIAN*" for November 12th, 1897 :—*

Probably the earliest discovery of cohesion under electric influence was contained in that old, forgotten observation of Guitard in 1850, that when dusty air was electrified from a point the dust particles tended to cohere into strings or flakes. The same thing no doubt occurs in the formation of snowflakes under the influence of atmospheric electrification ; and the cohesion of small drops into large ones in the proximity of a charged cloud is exceedingly familiar, since it results in the ordinary thunder-shower. Great light was thrown on these meteorological phenomena by the discovery of Lord Rayleigh in 1879 of the curious behaviour of a small fountain or vertical water-jet when exposed to the neighbourhood of a stick of excited sealing-wax. A smooth orifice being arranged to throw a jet of water about three or four feet nearly vertically, the jet breaks into drops, and the drops scatter in all directions, rebounding from one another and giving a shower of fine spray ; but if a stick of sealing-wax be rubbed on the sleeve of a coat and brought within one or two yards of the place where the jet breaks into drops, it will be found that the scattering ceases, the fine spray is no longer formed, and the broken jet rises and descends in great blobs of water. The rain-shower has, in fact, been converted into a thunder-shower. Further experiments, conducted chiefly with two jets, elucidated the phenomenon.* Arranging two nearly parallel jets from neighbouring orifices so as to impinge against each other, they were found ordinarily to rebound after colliding, a sort of film or superficial layer appearing to prevent amalgamation of the jets into one ; but if a slight difference of electric potential were maintained between the two jets, say by connecting them to the terminals of a Leclanché cell, then the boundary layer broke down,—the two colliding jets no longer separated with a rebound, but amalgamated and became one.

Lord Rayleigh developed a similar explanation for the single jet. The scattering of the jet in its ordinary state was due to the rebound of colliding drops, as could be *seen* by examining it with a sufficiently instantaneous or intermittent mode of illumination ; but if an electric charge were in the neighbourhood it must be supposed that a trace of potential difference existed between the drops, which caused them to amalgamate into one whenever they collided, and thus speedily to become united into a comparatively few large drops, which then continued on their parabolic way.

At first sight it would seem as if the neighbourhood of a negative charge should charge all the drops positively at the place whence they break off from the earth-connected parent jet, and should thus cause them all to repel each other. And if the electrified sealing-wax is held too close, this is exactly what happens. All the drops are then similarly electrified, and scatter more violently than ever, never in that case coming into any rebounding or other contact with each other. But under a gentler electric influence the similar charging has a less marked result, and a polarisation difference of potential of one or two volts may without difficulty be supposed to exist in the air between d.ops, partly because they are not all equally charged and partly because each is a conductor acted on inductively by a neighbouring electrified body. In this connection it must be remembered that rubbed sealing wax is at a potential of several thousand volts, and therefore can readily cause a potential gradient of two or three volts per millimetre throughout a yard or two of space.

---

* *Proc.* Roy. Soc., 1879 and 1882.

The next stage was the re-discovery, in 1883, of Guitard's old dust phenomenon by the present writer and the late J. W. Clark (*Nature*, July, 1883 ; *Phil. Mag.*, March, 1884), when they were working together at the dust-free region seen over hot bodies when strongly illuminated in dusty air. The fact of such dust-free spaces was discovered by Tyndall, and they can readily be seen by placing a lighted spirit lamp or a hot poker in the beam of an electric lamp. Tyndall thought the dust was calcined or burnt up, and that thus the air was freed from it ; but this is an utterly erroneous explanation, and the true explanation is of a more recondite character, being connected with the bombarding effect of gas molecules as illustrated in the Crookes radiometer. The dust particles are beaten away from the hot body by a molecular bombardment, which manifests itself even at ordinary pressures on bodies of sufficiently small size, as indeed was also otherwise shown by Tait and Dewar and Osborne Reynolds in the course of remarkable theoretical and practical investigations.*

Before arriving at this explanation, however, we experimented to see if the phenomenon was caused by the air having become slightly electrified, perhaps by reason of its having streamed as an upward convection current over the surface of the warm solid, at which we were looking, in a thick smoky atmosphere, in the concentrated light of an electric arc. We therefore purposely electrified the rod, to see what that would do, and we found to our surprise, directly the electric machine was turned, that the smoky atmosphere almost instantaneously disappeared, and the box became quite clear.

This experiment, after development, though described in July, 1883, was shown in public for the first time at the Dublin Royal Society (*Nature*, April 24, 1884), and subsequently at the British Association meeting in Montreal† in 1884, and was applied to the experimental clearing of rooms from dense smoke or fume. It has often been shown since, by Mr. Swan and others, and has become fairly well known.‡

---

* "Dimensional Properties of Matter," *Phil. Trans.*, 1879.

† Evening Lecture on "Dust," by the writer, see *Nature*, Vol. 31, p. 265 ; also *Journal* of the Royal Institution, May, 1886.

‡ Apparatus for the purpose is now in the catalogue of Messrs. Ducretet, of Paris, but they supply a pair of combs of points. It makes a more interesting experiment if only one point is used, in a moderate space, and

The next observation of cohesion under electrical influence was made by the writer in 1889, while working at the protection of telegraphic instruments and cables from lightning,—a research which resulted in the use of choke coils as supplementary to the air-gaps of the ordinary lightning guard, and thus to the forms of instrument constructed by Dr. Alex. Muirhead for telegraphic work in this country, and to the supplementary additions adopted by the Westinghouse Company for their non-arcing guards adapted to electric light and power installations in America. The observation of cohesion was a bye-issue, noticed when the knobs of the lightning guard were brought too close together.*

When lightning itself strikes a guard, it has indeed often been found that the opposite sides of the protective air-gap are fused together. This, no doubt, may be partly due to a straightforward melting or welding by heat, but it is probably not solely that. Molten metals without a flux do not so readily weld. It is almost certainly due to a cohesive action also, the difference of potential between the molten terminals resulting in adhesion and amalgamation, a phenomenon also observed in the frequent locking of an electric arc formed between two metallic electrodes. However this may be, certainly the phenomenon occurs on a small scale, for if the pair of knobs or points placed as a shunt to protect a galvanometer or other telegraphic instrument from lightning (or what is easier experimentally and essentially the same thing, from a Leyden jar discharge) be set too close together, the galvanometer will be found to be short-circuited after a spark, and the knobs will be found, both by mechanical and electrical tests, to be feebly united at a single point.† Not only, however, is the galvanometer short-circuited by the metallic junction so formed, but at the instant of the formation of the joint it experiences a very perceptible kick, indicating a momentary current, coincident no doubt with the electric discharge, but

---

the electric supply regulated so as not to hurry the disappearance of the smoke too quickly, but to exhibit the stages of aggregation which precede the final disappearance by deposition. Any kind of smoke serves, but a bit of magnesium ribbon burnt under a bell jar is cleanly and effective. It should be looked at in a window or other good light, of course.

* *Journal* of the Institution of Electrical Engineers for 1890, pp. 352-4.

† "Modern Views," second edition, p. 359.

one from which it would have been protected had not the junction occurred. The galvanometer kick is clearly an effect due to the uniting metals, but it has not yet been fully elucidated ; it seems to have been first observed by Mr. Stroh in his excellent researches on microphonic action, related in the *Journal* of the Society of Telegraph Engineers, 1883 and 1887, and it may possibly be thermo-electric, as Prof. Hughes, who also observed it, thinks likely; but it may be electro-chemical, or it may be connected with an effect observed later by FitzGerald in his galvanometer mode of detecting Hertzian waves, which he published at the Royal Institution in 1890. The point of present interest is the cohesion which sets in between the knobs when the spark occurs : an extremely feeble spark was found sufficient to produce the effect, provided the surfaces were already almost infinitely close together, *i.e.*, provided they were already in what would be called contact, with the merest imperceptible film of (probably) oxide separating them, just the kind of film which a chemical flux is useful in removing. The electrical stimulus appears to act as such a flux, and the adhesion of the two surfaces was demonstrated by an electric bell and single cell in circuit. Every time the spark occurred the bell rang, and continued ringing, until the table, or some part of the support of the knobs, was tapped so as to shake or jar them asunder again.* The arrangement constitutes a convenient detector in the syntonic Leyden jar experiment, depicted in Fig. 4, p. 6 (*see* also p. 21).

If the electric bell stands on the same table as the support of the sparking knobs, or, still better, if it be put into mechanical contact with them, its tremor is quite sufficient to break the contact asunder again ; unless the spark, and therefore the adhesion, has been too strong. Raising the bell into the air, it ceases to interrupt the spark-induced continuity, and in that case continues to ring ; but directly it is replaced so that its vibration can reach the cohered surfaces through their solid supports it usually happens that a few strokes—often, indeed, the first stroke—of the bell, sometimes even the incipient movement of the hammer preparatory to a stroke, is

---

* *Journal* of the Institution of Electrical Engineers, 1890, p. 352. *See* also remarks by Mr. Stroh in two microphone discussions, *Journal* of the Institution of Electrical Engineers, 1883 and 1887.

sufficient to break the circuit and suspend instantly the action, restoring the gap to its original condition and leaving the circuit ready to be completed again by another spark.

The spark in these early experiments was usually supplied from the outer coats of a pair of oppositely-charged small Leyden jars, whose knobs sparked into each other ; the idea being to ascertain all the conditions pertaining to the feeble residue of a lighting discharge which is liable to be conducted by tele-graph wires to a distance, and there cause some damage to sensitive instruments not suitably protected from sudden electric jerks, whose laws of flow are quite different from those proper to steady currents.

Meanwhile, in 1887 and 1888, had been performed the great experiments of Hertz on electric waves in free space. The writer, assisted by Prof. Chattock, had also made some experiments concerning the production and detection of waves on a system of long parallel wires stretched on insulators across and around a large room, and excited by the discharge of a pair of condensers, an arrangement very similar to that now known under the name of Lecher ; and clear experimental evidence of the existence of nodes and loops on such wires, as well as a method of approximately measuring the wave-length, was given.* The brush luminosity of the wires, afterwards observed more strikingly by Tesla, was also seen and shown to the Physical Society. The interest of these experiments was, however, altogether eclipsed by the brilliant and masterly investigation at Carlsruhe by Hertz, who, as everyone except the British public is aware, put into practice FitzGerald's 1883 suggestion that Leyden jar discharges should emit Max-wellian radiation, and conclusively demonstrated the existence and some of the properties of such waves by this very means ; using, however, Leydens of small capacity, and with the coat-ings well separated, so that the electrostatic energy of the charge should have an intensity comparable with the magnetic energy of the discharge, even at some distance from the circuit.

The whole subject of electric waves was thus laid open to physicists, and many have been the workers in the field.

---

* Verbally to Section A at Bath, 1888. *See* also *Phil. Mag.*, August, 1888, p. 229 ; and *The Electrician*, Vol. 21, pp. 607-8.

Trouton, of Dublin, worked long and successfully at their optical analogies, with the very inadequate means of detection then known;* and since better means have been known perhaps the most complete set of experiments published, after Hertz himself, is that contained in the book " Optice Elettrica," by Prof. Righi, of Bologna ; but some account of several previous researches is contained in the second edition of " Modern Views of Electricity," in the chapter called " Recent Progress," of date 1892. The means used by Hertz and his immediate followers to detect the waves was simply the little spark which they excited in conductors upon which they fell ; electric currents being set up in such conductors by the act of reflection. The effect was often at that time attributed to electric resonance or syntony, but there was very little true resonance in these experiments ; the first swing was usually much more powerful than any of the succeeding ones, and was competent to cause the little spark ; if it failed the remainder of the swings had but a poor chance of success. Consequently precision of tuning was not really important, though no doubt it would help a little.

It is interesting to note that a magnetic needle detector not unlike Rutherford's had been used long ago by Joseph Henry at Washington, and that minute induced sparks, identical in all respects with those discovered by Hertz, had been seen in recent times both by Edison and by Silvanus Thompson, being styled " etheric force " by the former ; but their theoretic significance had not been perceived, and they were somewhat sceptically regarded. Yet Henry, even in those pre-Maxwellian days, was led to an intuition concerning the spread of electrical disturbance surprisingly near the truth. The truth indeed it was in some sort, but it was not worked out or grasped in detail, and so cannot be considered as more than a brilliant guess ; but the fact that an observation of the widespread surgings induced in the neighbourhood of a primary discharge nad been made by Henry, and had been seen by others to be capable of giving actual sparks, before the time of Hertz, although it has no real bearing on Hertz's fresh discovery, and did not lead those who, like the writer, had long been trying to think of a detector for Maxwellian waves to discover one, never-

* *Nature,* Vols. 39 and 40.

theless is instructive as showing how frequently it happens that a fact is lying ready to hand but is not taken up and appreciated until some special or extra stimulus has been supplied.

After Hertz's results had become well-known, the writer devised a plan whereby real electric resonance could be demonstrated with a pair of actual glass Leyden jars of ordinary pattern, by connecting each to a discharge circuit, the one complete, the other with an air-gap, and providing the first or receiving jar with an overflow path or bye-circuit provided with an air-gap across which a visible spark could occur whenever the induced oscillations or surgings accumulated in its main circuit were sufficiently intense to make the jar overflow.* The air-gap was most easily provided by a strip of tinfoil pasted over the lip of the jar, but it served equally well if wires led from the two coatings to a pair of adjustable knobs near together, like a lightning guard, between which the overflow spark could pass. The same knobs indeed were used as had already served for the lightning experiments; and, as in that case, if the knobs are arranged very close together and are put in circuit with a battery and a bell, cohesion sets in and the bell rings whenever the overflow occurs. The bell continues to ring until the stand is tapped, but if the bell itself touches the stand or the table, it rapidly breaks contact by its vibration, exactly as described, p. 77 (*see* also Fig. 16A, p. 21). Closed Leyden jar circuits are not strong radiators, nor was this resonance then observed excited by true waves. No attempt was at this time made to apply the cohesion principle to the detection of true Hertz waves such as could be felt at a considerable distance from a strongly radiating source.

Before this time, FitzGerald and Trouton had hit upon their galvanometer method of demonstrating to an audience the occurrence of the minute scarcely-visible spark in the gap of a Hertz receiver.†

Prof. Minchin also, working at Cooper's-hill with his sensitive photo-electric cells, especially with some which he called

---

* *Nature*, Vol. 41, p. 363 ; or, "Modern Views of Electricity," second edition, p. 338. *See* also Fig. 4, p. 6.

† *Nature*, Vol. 41, p. 295 ; and Vol. 42, p. 172.

"impulsion cells," that behaved abnormally when subjected to taps or other mechanical vibrations, found that when Mr. Gregory was working a Hertz radiator in another part of the same laboratory the electrometer connected to his cells responded.* Many other detectors have been devised and used, but this of Minchin's almost certainly depends on the cohesion principle, though its action seemed paradoxical then.   Moreover he was able, by its aid, to signal without wires over a considerable number of yards, at that early date (1890 and 1891).

About the same time, Prof. Boltzmann used a charged goldleaf electroscope for the same purpose, having it so arranged that the electroscope was on the point of discharging across a minute air-gap, so that its leaves were dilated by a definite amount.   The slightest excess of charge would make it discharge and the leaves instantly collapse.   In this charged condition it was sensitive to very minute electric surgings, and if Hertz waves were excited in another part of the room, the wave disturbances caused the gap to break down and the electroscope leaves to collapse.† This method is not a cohesion method, but it led the writer, when subsequently repeating Boltzmann's results with modifications, to realise that, if the gap were almost closed, cohesion could be made to set in by the surgings induced by regular Hertz waves (Fig. 16, p. 18).

The Boltzmann gap method was accordingly modified in several ways ; one way was to make it of carbon and to connect it, with its wave collector, to the terminals of 110-volt electric light leads, so that whenever a Hertz vibrator was discharged and induced a minute spark across the gap, that same spark might close the circuit and establish an arc.   This plan forced itself on my attention by the behaviour of sundry Swan lamps suspended with shades so as to illuminate my lecture table, which became short-circuited whenever a large Hertz vibrator was at work ; for the lamps were at that time kept from rotation, and thereby from glaring into the eyes of the audience instead of being screened from them, by a couple of copper wires stretched across the theatre.   So long as those wires were there, the fuses used to blow whenever a Hertz oscillator was started ; an experiment which was interesting enough, and was

---

* *Phil. Mag.*, March, 1891 ; also January, 1894
† *Wied. Ann.*, Vol. 40.

shown to several people, including, I think, Prof. FitzGerald, but which was sufficiently a nuisance to necessitate the wires, which were acting as collecting wires, being taken down and replaced by stretched silk threads, which are there to this day. Another modification was to connect the gap to an Abel's fuse or to a gas leak, which exploded or ignited under the influence of a feeble spark. Yet another was to connect it to a single cell and electric bell or galvanometer, as already explained.

Meanwhile, however, and well before these later experiments on the detection of Hertz waves were in progress, certain discoveries had been made by M. Branly, Professor of Physics in the Catholic Institute of Paris, which were of the greatest interest and importance. Prof. Branly had found that a coat or varnish of fine copper dust, porphyrised copper or other such substance, though it could only conduct a current very feebly, and much as a blacklead pencil trace conducts, under ordinary conditions, yet fell in resistance enormously whenever an electric spark occurred in its neighbourhood; somewhat in the fashion that the resistance of selenium falls on exposure to light. It is not clear that M. Branly recognised that he was dealing with Hertz waves or true electrical radiation, but his observations were most satisfactory and conclusive, and he measured the reduction of resistance caused in a number of different substances, including an assemblage of metallic filings, and conglomerates or paste of filings in various viscous liquids and in dry powders. Moreover, he found that the spark was still operative in reducing resistance even when it was several yards distant.

The account of Prof. Branly's experiments is to be found in a couple of short communications to the French Academy of Science (*Comptes Rendus*, Vols. 111 and 112), and the writer had intended to reproduce in abstract the gist of these memoirs; but to readers of *The Electrician* this is unnecessary, as a descriptive article from *La Lumière Electrique* has already been translated in full, in July and August, 1891 (see *The Electrician*, Vol. XXVII., pp. 221 and 448, now reproduced as Appendix). Unfortunately the writer, in common perhaps with others, must confess to having overlooked these articles at the time, probably by reason of their coincidence with the holiday season. In his second edition of "Modern Views of Electricity," published in 1892, though he refers on page 359 to the cohesion principle in this connection, the writer is clearly ignorant of Branly's experiments.

The matter seems to have been ignored in this country till 1892, when Dr. Dawson Turner described the experiments to the British Association in Edinburgh, and even till 1893, when Mr. Croft brought them to the notice of the London Physical Society. Prof. Minchin at once realised that here was a phenomenon analogous to what he had been observing with his impulsion cells, and after a few trials wrote a Paper to the Physical Society recounting his repetitions and modifications of Branly's experiments.* This Paper, before it was read, was circulated by the Society to its country members, and so came to the eye of the writer, who at once wrote a short note summarising some of his work in the same direction, and pointing out that this discovery of Branly's, thus made known to him, was another case of the electrical cohesion phenomenon already observed by several experimenters. This is published along with Prof. Minchin's Paper in the *Phil. Mag.* for January, 1894, and to it the friendly reader is referred. The writer at once proceeded to try the Branly tube of filings, and found it far superior in manageability to either the Boltzmann gap or his own delicately adjusted cohering knobs; though immediately afterwards he and FitzGerald together arranged a single point coherer, of iron and aluminium (point of sewing needle resting on aluminium foil), of what was at that time extraordinary sensitiveness and of reasonable manageability. A whole series of quasi-optical experiments were then undertaken with the new detector, and were shown to students and to the Liverpool Physical Society; moreover, before long, various improved methods of arranging the filings were gradually adopted, especially by sealing them up in vacuum or in an atmosphere of hydrogen (*see* page 34) so as to protect them from continued oxidation by the air, and to prevent the film which hypothetically separates the surfaces from growing too thick. Indeed, brass filings in hydrogen speedily got *too* clean, and became so sensitive that it was almost impossible to restore the original high resistance by

---

* *Phil. Mag.*, January, 1894.

tapping. Consequently, a perfect or Sprengel vacuum was preferred to hydrogen. Almost any filings tube could detect signals from a distance of 60 yards, with a mere six-inch sphere as emitter and without the slightest trouble, but the single-point coherer was usually much more sensitive than any filings tube. Mr. Shelford Bidwell has also worked with varieties of powder.

The tapping back was at first performed by hand, and for optical experiments this is still, perhaps, the most convenient plan; but automatic tappers were very soon arranged, just as with the old knobs; an electric bell mounted on the base of a filings tube (*see* page 31) was not found very satisfactory, however, because of the disturbances caused by the little sparks at its contact breaker, to which the pre-vious coarser knob-arrangements had failed to respond; so a clockwork tapper, consisting of a rotating spoke wheel driven by the clockwork of a Morse instrument, and giving to the filings tube or to a coherer a series of jerks occurring at regular intervals, to imitate what the writer supposed must occur in the eye, viz., a restoration to sensitiveness after an interval corresponding to the persistence of impression, was also employed. Many of these things were shown at a Friday evening lecture at the Royal Institution on June 1, 1894, while others were shown the same autumn at the B.A. meeting at Oxford. In both cases signalling was easily carried on from a distance through walls and other obstacles, an emitter being outside and a galvanometer detector inside the room. Distance without obstacle was no difficulty in these experiments, only free distance is not very easy to get in a town, and stupidly enough no attempt was made to apply any but the feeblest power so as to test how far the disturbance could really be detected. Mr. Rutherford, however, with a magnetic detector of his own invention, constructed on a totally different principle, and probably much less sensitive than a coherer, did make the attempt and succeeded in signalling across half a mile, full of intervening streets and houses at Cambridge.*

Numbers of people have worked at the detection of Hertz waves with filing tube receivers, and every one of them must have known that the transmission of telegraphic messages in

---

* *Phil. Trans.*, 1897, A., communicated to the Royal Soc., June, 1896,

this way over moderate distances was but a matter of demand and supply; Sir W. Crookes, indeed, had already clearly stated this telegraphic application of Hertz waves in the *Fortnightly Review* for February, 1892, and refers to certain experiments already conducted in that direction,* the details of which are unknown to the writer (but see Appendix I.). There remained no doubt a number of points of detail, and considerable improvements in construction, if the method was ever to become practically useful; but these details could safely be left to those who had charge of the Government monopoly of telegraphs, especially as their eminent Head was known to be interested in this kind of subject.

Meanwhile the optical developments of the matter excited most interest among physicists, both here and on the continent; the writer performed some experiments of the kind, Prof. Righi at Bologna performed many more, and Prof. Chunder Bose, of Calcutta, repeated several of them with additions and improvements, using as detector a sort of half-way house between a point coherer and a filings tube by squeezing a few rolls or spirals of wire between a point and a micrometer screw. Restoration to sensitiveness was in this case achieved by relaxing the pressure of the screw, and the writer has not found Bose's form of coherer specially convenient; but Prof. Bose's whole apparatus, constructed as it was precisely on lines published by the writer, was well designed in detail and exceedingly compact, being on the scale of an ordinary goniometer; and with it many experiments familiar in ordinary optics could readily be shown with electric radiation.

In all the optical experiments made by any of these observers it was customary to place the axis of the emitter either horizontally, or vertically, or inclined, in other words to emit radiation polarised in any azimuth (or rather altitude), and to arrange the collecting part of the receiver to correspond or otherwise, according as response or no response was desired. In fact, observations on polarisation were the easiest and the most instructive that could be made with the definite kind of radiation now for the first time at command.

---

* Quoted in *The Electrician* "Notes," October 1, 1897

The rotation of the plane of polarisation, the conversion of plane into elliptical polarisation, the amount of radiation reflected by substances at different angles and different aspects with regard to the direction of vibration, were readily observed. Furthermore, ever since Hertz's first discovery, whenever waves had to travel through a metal grid or alongside a plane conductor, it was natural to arrange the electric oscillations so as to be normal to the conducting lines or plane, for if they were tangential they excited electric currents therein, and their energy became wasted in the production of heat. So, in so far as earth and water are conductors, it is desirable to use radiation polarised in a horizontal plane, *i.e.*, with the electric oscillations vertical, if considerable distances are to be traversed by it.

With respect to an explanation *why* metallic cohesion is caused under electrical influence, the following considerations are offered :—

Mr. Rollo Appleyard made a liquid coherer of two globules or pools of mercury, side by side and touching, but kept apart by a thin film of grease, such as is easily given by a coat of paraffin oil. Connecting up a battery cell to these mercury pools through a key, he found that every time the key is depressed the pools move together and become one ; he points out moreover that mercury globules shoot out a tentacle towards the positive terminal (on the principle of the capillary electrometer, of course), and this must be taken into account in any coherer theory.* Lord Rayleigh also devised and exhibited a liquid form of coherer. It is interesting to observe, as he points out, that in a mercury form of coherer an appreciable time interval occurs between the depression of the key and the amalgamation of the mercury, the lag looking as if a film had to be mechanically squeezed out between the oppositely-charged mercury surfaces, and as if this took a perceptible fraction of a second to accomplish. This experiment conveys the useful suggestion that cohesion may in all cases be the result of electrostatic attraction, and that the molecular films separating solids in contact may thus also have to be squeezed out, though as they only touch at single points such

---

* *Phil. Mag.*, May and July, 1897. He also shows electrical cohesion by an emulsion of oil and water, the two liquids, thoroughly shaken up, at once separating when exposed to strong electrical influence.

extrusion is almost instantaneously achieved. This may very likely be the chief cause, for although a true electro-chemical extension of the range of cohesion between polarised molecules had seemed to the writer to be a possible explanation also, he now perceives that the electrostatic force alone may be sufficient. For it is easy to calculate the force of attraction between two surfaces differing in potential by a volt, and separated from one another by the smallest known thickness of thin film (which is 10^{-7} centimetre, or 1 millimicron, called $\mu\mu$ by microscopists); such force per unit area would be given by the square of the potential gradient divided by 8π, that is, it would amount to $\dfrac{1}{25}\left(\dfrac{10^7}{300}\right)^2$ dynes per square centimetre, which equals 44 atmospheres, and is a very considerable pressure. A hundred times this attractive pressure would exist if the surfaces were within really *molecular* distance of each other; in addition to the force of true cohesion which would then, still more powerfully, operate; but the film thickness assumed above is such as would just prevent the force of cohesion from effectively acting across the gap, and would leave the electrical attraction due to the one volt alone. Three and a-half volts could therefore squeeze metals together with a force equal to a ton load per square inch, and might thus be sufficient to cause them to weld or unite, especially if the electric stimulus simultaneously acted in any way as a flux, by reducing the infinitesimal tarnish of oxide or other compound which must be supposed normally to cover them.

In so far as the approximate contact is not between *surfaces*, but between points consisting of relatively few molecules, the attractive pressure is greater rather than less. Thus to take an extreme case, the attraction between two oppositely-charged molecules differing only by a volt from each other, and separated by a thin film like the black spot of a soap-film whose thickness was so admirably measured by Profs. Reinold and Rücker, is over 1,000 atmospheres in intensity. These differences of potential across thin films cannot continue for any time, unless a battery is used, for the films do not really insulate; they are able however to act as dielectrics for an instant, and to be burst with what we must be allowed to call a spark, though an infinitesimally small one, if the momentary strain caused by the impulsive rush of electricity is too great

APPENDIX I.

PROF. HUGHES'S OBSERVATIONS.

An account of the history of the coherer principle would
not be complete without a reference to an interesting remi-
niscence of early observations recently put on record by the
discoverer of the microphone. At each stage of his observations of electrical cohesion between metals the author was
confronted by a reference to some earlier observations of Prof.
Hughes, and he felt sure that during the work on the microphone many or all of the phenomena he was then observing
must have been previously encountered by Prof. Hughes.
No full account was at that time available, however, but now
it is clear that the observations were made (like some of
Edison's on what he called etheric force, and like the
very remarkable still earlier ones of Joseph Henry) before
the time of Hertz, when the existence of electric waves able
to excite sparks or perform other energetic acts was unlooked
for and not clearly understood.

Nevertheless, at this early period it is clear that Prof.
Hughes observed, though he did not follow up the observation,
not only the occurrence of electric waves or impulses in space,
but also the coherer method of detecting them ; in fact, that
he unwittingly made the earliest experiments on wireless
telegraphy by this plan.

The simplest way is to quote Prof. Hughes's letter to Mr.
J. J. Fahie from *The Electrician*, May 5, 1899, p. 40,
beginning with Mr. Fahie's letter as an introduction :—

Extract from recent letter from Mr. J. J. Fahie to Prof. Hughes.

"DEAR PROF. HUGHES : I have now in the press a history of
Wireless Telegraphy from 1838 to 1899, and in writing to Sir
William Crookes for information he tells me that many years

ago he saw some experiments of yours with the microphone, in which you signalled from one part of a house to another without connecting wires, and he desires me to refer to you for particulars. I think, with Sir William, that it is a pity you have not hitherto published your results, and I sincerely hope you will now do so. If also you would kindly favour me with a short account, I could find room for it in my book, which is now in the printer's hands.—Sincerely yours,

J. J. FAHIE.

" Claremont Hill, St. Helier's, Jersey, April 26, 1899."

Reply from Prof. D. E. Hughes :—

40, Langham-street, W., April 29, 1899.

DEAR SIR : In reply to yours of the 26th inst., in which you say that Sir William Crookes has told you " that he saw some experiments of mine on aërial telegraphy, in about December, 1879, of which he thinks I ought to have published an account," and of which you ask for some information, I beg to reply with a few leading experiments that I made on this subject from 1879 up to 1886 :—

" In 1879, being engaged upon experiments with my microphone, together with my induction balance, I remarked that at some time I could not get a perfect balance in the induction balance, through apparent want of insulation in the coils, but investigation showed me that the real cause was some loose contact or microphonic joint excited in some portion of the circuit. I then applied the microphone in the circuit, and found that it gave a current or sound in the telephone receiver, no matter if the microphone was placed direct in the circuit, or placed independently at several feet distance from the coils, through which an intermittent current was passing. After numerous experiments, I found that the effect was entirely caused by the extra current produced in the primary coil of the induction balance.

" Further researches proved that an interrupted current, in any coil, through which an electric current was sent, gave out at each interruption of the primary current, such intense extra currents, that the whole atmosphere in the room (or in several rooms distant) would have a momentary invisible charge, which became evident if a microphonic joint was used as a receiver to a telephone.   This led me to experiment upon the best form of a receiver for these invisible electric waves, which evidently permeated great distances, and through all apparent obstacles, such as walls, &c.   I found that all microphonic contacts or joints were extremely sensitive. Those formed of a hard carbon such as coke, or a combination of a piece of coke resting upon a bright steel contact, were very sensitive and self-restoring ; whilst a loose contact between metals was equally sensitive, but would cohere, or remain in full contact, after the passage of an electric wave.

" The sensitiveness of these microphonic contacts in metals has since been rediscovered by Mons. Ed. Branly, of Paris, and by Prof. Oliver Lodge, in England, by whom the name of ' coherer ' has been given to this organ of reception ; but, as we wish this organ to make a momentary contact and not cohere permanently, the name seems to me ill-suited for the instrument.   The most sensitive and perfect receiver that I have yet made does not cohere permanently, but recovers its original state instantly, and, therefore, requires no tapping or mechanical aid to the separation of the contacts after momentarily being brought into close union.

" I soon found that, whilst an invisible spark would produce a thermo-electric current in the microphonic contacts (sufficient to be heard in the telephone in its circuit), it was far better and more powerful to use a feeble voltaic cell in the receiving circuit, the microphonic joint then acting as a relay, by increasing and diminishing the resistance at the contact, by the influence of the electric wave received through the atmosphere.

" I will not describe the numerous forms of the transmitter, and receiver, that I made in 1879, all of which I wrote down in several volumes of manuscripts in 1879 (but these have never been published), most of which can be seen here at my residence at any time ; but I will confine myself now to a few salient points.   I found that very sudden electric impulses, whether given out to the atmosphere through the extra current from a coil, or from a frictional electric machine, equally affected the microphonic joint, the effect depending more on the sudden high potential effect than any prolonged action.   Thus, a spark obtained by rubbing a piece of sealing-wax was equally as effective as a discharge from a Leyden jar battery of the same potential. The rubbed sealing-wax, or charged Leyden jar, had no effect, until they were discharged by a spark,—and it was evident that this spark, however feeble, acted upon the whole surrounding atmosphere in the form of waves, or invisible rays, of which I could not at the time determine. Hertz, however, by a series of original and masterly experiments, proved in 1887–9, that they were real waves similar to light, but of a lower frequency, though of the same velocity. In 1879, whilst making these experiments on aërial transmission, I had two different problems to solve: 1st, What was the true nature of these electric aërial waves, which seemed, whilst not visible, to spurn all idea of insulation, and to permeate all space to a distance undetermined. 2nd, To discover the best receiver that could act upon a telephone or telegraph instrument, so as to be able to utilise (when required) these waves for the transmission of messages. The second problem came easy to me, when I found that the microphone, which I had previously discovered in 1877–8, had alone the power of rendering these invisible waves evident, either in a telephone or galvanometer, and up to the present time I do not know of anything approaching the sensitiveness of a microphonic joint as a receiver. Branly's tube, now used by Marconi, was described in my first Paper to the Royal Society (May 8, 1878), as the microphone tube, filled with loose filings of zinc and silver, and Prof. Lodge's coherer is an ordinary steel microphone, used for a different purpose from that in which I first described it.

"During the long-continued experiments on this subject, between 1879 and 1886, many curious phenomena came out which would be too long to describe. I found that the effect of the extra current in a coil was not increased by having an iron core as an electromagnet—the extra current was less rapid and, therefore, less effective. A similar effect of a delay was produced by Leyden jar discharges. The material of the contact-breaker of the primary current had also a great effect. Thus, if the current was broken between two, or one, piece of carbon, no effect could be perceived of aërial waves, even at short distances of a few feet. The extra current from a small coil, without iron, was as powerful as an intense spark from a secondary coil, and at that time my experiments seemed to be confined to the use of a single coil of my induction balance, charged by six Daniell cells. With higher battery power, the extra current invariably destroyed the insulation of the coils.

" In December, 1879, I invited several persons to see the results then obtained. Amongst others who called on me and saw my results were :—

" December, 1879.—Mr. W. H. Preece, F.R.S.; Sir William Crookes, F.R.S.; Sir W. Robert Austen, F.R.S.; Prof. W. Gryll Adams, F.R.S.; Mr. W. Groves.

" February 20, 1880.—Mr. Spottiswoode, Pres.R.S.; Prof. Huxley, F.R.S.; Sir George Gabriel Stokes, F.R.S.

"November 7, 1888.—Prof. Dewar, F.R.S.; Mr. Lennox, Royal Institution.

" They all saw experiments upon aërial transmission, as already described, by means of the extra current produced from a small coil and received upon a semi-metallic microphone, the results being heard upon a telephone in connection with the receiving microphone. The transmitter and receiver were in different rooms, about 60ft. apart. After trying successfully all distances allowed in my residence in Portland-street, my usual method was to put the transmitter in operation and walk up and down Great Portland-street with the receiver in my hand, with the telephone to the ear.

" The sounds seemed to slightly increase for a distance of 60 yards, then gradually diminish, until at 500 yards I could no longer with certainty hear the transmitted signals. What struck me as remarkable was that, opposite certain houses, I could hear better, whilst at others the signals could hardly be perceived. Hertz's discovery of nodal points in reflected waves (in 1887–9) has explained to me what was then considered a mystery.

" At Mr. A. Stroh's telegraph instrument manufactory, Mr. Stroh and myself could hear perfectly the currents transmitted from the third story to the basement, but I could not detect clear signals at my residence about a mile distant. The innumerable gas and water pipes intervening seemed to absorb or weaken too much the feeble transmitted extra currents from a small coil.

" The President of the Royal Society, Mr. Spottiswoode, together with the two hon. secretaries, Prof. Huxley and Prof. G. Stokes, called upon me on February 20, 1880, to see my experiments upon aërial transmission of signals. The experiments shown were most successful, and at first they seemed astonished at the results, but towards the close of three hours' experiments Prof. Stokes said, ' that all the results could be explained by known electromagnetic induction effects, and therefore he could not accept my view of actual aërial electric waves unknown up to that time, but thought I had quite enough original matter to form a Paper on the subject to be read at the Royal Society.'

" I was so discouraged at being unable to convince them of the truth of these aërial electric waves, that I actually refused to write a Paper on the subject, until I was better prepared to demonstrate the existence of these waves ; and I continued my experiments for some years, in hopes of arriving at a perfect scientific demonstration of the existence of aërial electric waves, produced by a spark from the extra currents in coils, or from frictional electricity or secondary coils. The triumphant demonstration of these waves was reserved to Prof. Hertz, who by his masterly researches upon the subject in 1887-9 completely demonstrating not only their existence but their identity with ordinary light, in having the power of being reflected and refracted, &c., with nodal points, by means of which the length of the waves could be measured, Hertz's experiments were far more conclusive than mine, although he used a much less effective receiver than the microphone or coherer.

" I then felt it was now too late to bring forward my previous experiments, and through not publishing my results, and means employed, I have been forced to see others remake the discoveries I had previously made as to the sensitiveness of the microphonic contact, and its useful employment as a receiver for electric aërial waves. Amongst the earliest workers in the field of aërial transmission I would draw attention to the experiments of Prof. Henry, who describes in his work, published by the Smithsonian Institute, Washington, D.C., U.S.A., Vol. I., p. 203 (date unknown, probably about 1850), that he magnetised a needle in a coil at 30ft. distance, and magnetised a needle by a discharge of lightning at eight miles distance.

"Marconi has lately demonstrated that by the use of the Hertzian waves and Branly's coherer he has been enabled to transmit and receive aërial electric waves to a greater distance than previously ever dreamed of by the numerous discoverers and inventors who have worked silently in this field. His efforts at demonstration merit the success he has received ; and if (as I have lately read) he has discovered the means of concentrating these waves on a single point desired without diminishing its power, then the world will be right in placing his name on the highest pinnacle in relation to aërial electric telegraphy.—Yours, &c., D. E. HUGHES."

APPENDIX II.

VARIATIONS OF CONDUCTIVITY UNDER ELECTRICAL INFLUENCE.

The following is abstracted from an article by M. E. Branly in *La Lumière Electrique* of May 16, 1891, and is taken from *The Electrician* of June 26, 1891 :—

The object of this article is to describe the first results obtained in an investigation of the variation of resistance of a large number of conductors under various electrical influences. The substances which up to the present have presented the greatest variations in conductivity are the powders or filings of metals. The enormous resistance offered by metal in a state of powder is well known ; indeed, if we take a somewhat long column of very fine metallic powder the passage of the current is completely stopped. The increase in the electrical conductivity by pressure of powdered conducting substances is well known, and has had various practical applications. The variations of conductivity, however, which occur on subjecting conducting bodies to various electrical influences have not been previously investigated.

The Effect of Electric Sparks.—Let us take a circuit comprising a single cell, a galvanometer, and some powdered metal enclosed in an ebonite tube of 1 square centimetre cross section and a few centimetres long. Close the extremities of the tube with two cylindrical copper tubes pressing against the powdered metal and connected to the rest of the circuit. If the powder is sufficiently fine, even a very sensitive galvanometer does not show any evidence of a current passing. The resistance is of the order of millions of ohms, although the same metal melted or under pressure would only offer (the dimensions being the same) a resistance equal to a

fraction of an ohm.   There being, therefore, no current in the
circuit, a Leyden jar is discharged at some little distance off,
and the abrupt and permanent deflection of the galvanometer
needle shows that an immediate and a permanent reduction of
the resistance has been caused.   The resistance of the metal
is no longer to be measured in millions of ohms, but in
hundreds.   Its conductivity increases with the number and
intensity of the sparks.

Some 20 or 30 centimetres from a circuit comprising some
metallic filings contained in an ebonite cup, let us place a
hollow brass sphere, 15 to 20 centimetres in diameter,
insulated by a vertical glass support.   The filings offer an
enormous resistance and the galvanometer needle remains at
zero.   But if we bring an electrified stick of resin near the
sphere, a little spark will pass between the stick and the
sphere, and immediately the needle of the galvanometer is
violently jerked and then remains permanently deflected.   On
some fresh filings being placed in the ebonite cup, the resistance of the circuit will again keep the needle at zero.   If now
the charged brass sphere is touched with the finger, there is a
minute discharge and the galvanometer needle is again
deflected.   With a few accumulators the experiment can
easily be made without a galvanometer.   The circuit consists
of the battery, some metallic powder, a platinum wire, and a
mercury cup.   The resistance of the powder is so high that
the interruption of the circuit takes place without any
sparking at the mercury cup.   If now a Leyden jar is
discharged in the neighbourhood of the circuit the powder is
rendered conducting, the platinum wire immediately becomes
red hot, and a violent spark occurs on breaking the circuit.

The influence of the spark decreases as the distance
increases, but its influence is observable several metres away
from the powder, even with a small Wimshurst machine.
Repeating the spark increases the conductivity ; in fact, with
certain substances successive sparks produce successive jerks,
and a gradually increasing and persistent deflection of the
galvanometer.

Influence of a Conductor traversed by Condenser Discharges.—
While using a Wimshurst machine it was noticed that the
reduction in the resistance of the filings frequently took place

before discharge. This led me to the following experiment: Take a long brass tube, one end of which is close to the circuit containing the metallic powder; its other end, several metres distant from the circuit, is fairly close to a charged Leyden jar. A spark takes place and the conductor is charged. At the same instant, the conductivity of the metallic powder is greatly increased.

The following arrangement, owing to its efficacy, convenience, and regularity of action was used by me in most of my researches, and I shall briefly call it the A arrangement (*see* Fig. 53).

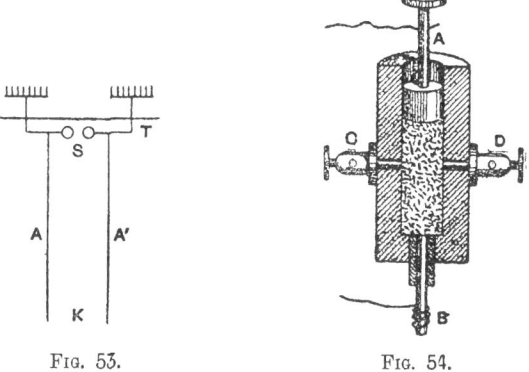

FIG. 53.   FIG. 54.

The source of electricity is a two-plate Holtz machine driven at from 100 to 400 revolutions. A sensitive substance is introduced into one of the arms of a Wheatstone bridge, or into the circuit of a single Daniell cell at a distance of some 10 metres (34ft.) from the Holtz machine. Between the discharge knobs of the machine and the Wheatstone bridge, and connected to the former, there are two insulated brass tubes, A A', running parallel to one another 40 centimetres apart. The Leyden jars usually attached to a Holtz machine may be dispensed with, the capacity of the long brass tubes being in some measure equivalent to them. The knobs S were 1mm., ·5mm., or ·1mm. apart. When the plates were rotated sparks rapidly succeeded each other. Experiments showed that these sparks had no direct effect at a distance of 10 metres. The two tubes A A' are not absolutely necessary,

H

the diminution of resistance is easily produced if only one is employed, and in some cases, indeed, a single conductor is more efficacious. An increase in the speed of the machine increases its action to a marked extent. The sparks at S may be suppressed by drawing the knobs apart, but the conductor A will still continue to exert its influence, especially if there is a spark gap anywhere about.

Effects of Induced Currents.—The passage of induced currents *through* a sensitive substance produces similar effects to those described above. In one instance an induction coil was taken, having two similar wires. The circuit of the secondary wire was closed through a tube containing filings, the galvanometer being also in circuit. Care was taken to ascertain before introducing the filings into the circuit that the currents on make and break gave equal and opposite deflections. Filings were then introduced into the circuit, the primary being made and broken at regular intervals. The following table gives the results obtained in the case of zinc filings :—

ZINC FILINGS.

Galvanometer throws.		Galvanometer throws.	
1st closing	1°	1st opening	18°
2nd ,,	64°	2nd ,,	100°
3rd ,,	146°	3rd ,,	140°

Effects of Passing Continuous Currents of High E.M.F.—If a continuous current of high E.M.F. is employed, it renders a sensitive substance conducting. The phenomenon may be shown in the following manner. A circuit is made up consisting of a battery, a sensitive substance, and a galvanometer. The E.M.F. of the battery is first one volt, then 100 volts, then one volt. Below I give the galvanometer deflections obtained with an E.M.F. of one volt for three different substances before and after the application of the E.M.F. of 100 volts :—

Before application of current.	After application of current.
16	100
0	15
1	500

In the case of some measurements taken on a Wheatstone bridge a prism of aluminium filings interposed between two copper electrodes offered a resistance of several million ohms before a high E.M.F. was applied, but only offered a resistance of 350 ohms after the application of this pressure for one minute. The time during which the powder should be interposed in the battery circuit should not be too short. Thus, in one instance, the application for 10sec. of 75 mercury sulphate cells produced no effect, but their application for 60sec. resulted in the resistance being reduced from several megohms to 2,500 ohms.

It should be observed that the phenomenon of suddenly increased conductivity occurs, even if the sensitive substance is not in circuit with a battery at the time it is influenced. Thus, the metallic filings, after having been placed in circuit with a Daniell cell, and its high resistance observed, may then be completely insulated and submitted in this condition to the action of a distant spark, or of a charged rod, or of induced currents. If, after this, the filings are replaced in their original circuit, the enormous increase in their conductivity is immediately apparent.

The conductivity produced by these various methods takes place throughout the whole mass of the metallic filings, and in every direction, as the following experiment will show. A vertical ebonite cup containing aluminium powder (*see* Fig. 54) is placed between two metal plates, A, B; laterally the powder is in contact with two short rods, C, D, which pass through the sides of the ebonite cylinder. A and B can be connected to two terminals of one of the arms of the Wheatstone bridge, C and D being free, and *vice versâ*. Whatever arrangement is adopted, if a battery of 100 cells is joined up for a few seconds with one or the other of the pairs of terminals, the increase in the conductivity is immediately visible in that direction, and is found to exist also in the direction at right angles.

Substances in which Diminution of Resistance has been Observed. —The substances in which the phenomenon of the sudden increase of conductivity is most easily observed are filings of iron, aluminium, copper, brass, antimony, tellurium, cadmium, zinc, bismuth, &c. The size of the grains and their nature are not the only elements to be considered, for grains of lead of the same size, but coming from different quarters, offer at the same temperature great differences in resistance (20,000 to 500,000 ohms). Extremely fine metallic powder, as a rule, offers almost perfect resistance to the passage of a current. But if we take a sufficiently short column and exert a sufficiently great pressure a point is soon reached when the electrical influence will effect a sudden increase in the conductivity. Thus, a layer of copper reduced by hydrogen, which does not become conducting under the influence of the electric spark or otherwise, will become so on being submitted to a pressure of 500 grammes to the square centimetre (7lb. per square inch). Instead of using pressure, I employed as a conductor in some experiments a very fine coating of powdered copper spread on a sheet of unpolished glass or ebonite E (Fig. 55), seven centimetres long and two centimetres broad.

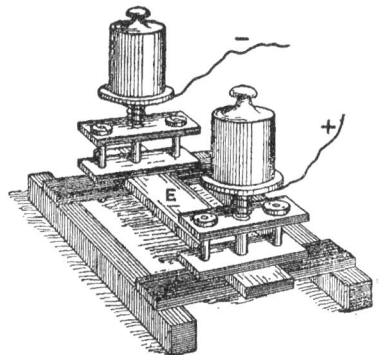

Fig. 55.

A layer of this kind, polished with a burnisher, has a very variable resistance. With a little care one can prepare sheets which are more or less sensitive to electrical action.

Metal powder or metal filings are not the only sensitive substances, as powdered galena, which is slightly conducting under pressure, conducts much better after having been submitted to electrical influence. Powdered binoxide of maganese is not very sensitive unless mixed with powdered antimony and compressed.

Making use of the A arrangement, with very short sparks at S (Fig. 35), the phenomenon of increased conductivity can be observed with platinised and silvered glass, also with glass covered with gold, silver and aluminium foil. Some of the mixtures employed had the consistency of paste. These were mixtures of colza oil and iron, or antimony filings, and of ether or petroleum and aluminium, and plumbago, &c. Other mixtures were solid. If we make a mixture of iron filings and Canada balsam, melted in a water bath, and pour the paste into a little ebonite cup, the ends of which are closed by metallic rods, a substance is obtained which solidifies on cooling. The resistance of such a mixture is lowered from several megohms to a few hundred ohms by an electric spark. Similar results are obtained with a solid rod composed of fused flowers of sulphur and iron or aluminium filings, also by a mixture of melted resin and aluminium filings. In the preparation of these solid sensitive mixtures care must be taken that the insulating substance should only form a small percentage of the whole.

Some interesting results are also obtained with mixtures of sulphur and aluminium, and with resin and aluminium, when in a state of powder. When cold, these mixtures as a rule do not conduct either directly or after they have been exposed to electrical influences, but they become conducting on combining pressure with electrical influences. Thus, a mixture of flowers of sulphur and aluminium filings in equal volumes was placed in a glass tube 24mm. in diameter. The weight of the mixture was 20 grammes, and the height of the column 22mm. ; with a pressure of 186 grammes per square centimetre ($2\frac{1}{2}$lb. per square inch). The mixture is not conducting, but after exposure to electrical influence, obtained by the A arrangement, the resistance falls to 90 ohms. In a similar manner a mixture of selenium and aluminium, placed in a tube 99mm. long, was not conducting until after it was exposed to the combined influence of pressure and electricity.

The following is one of the group of numerous experiments of a slightly different character. A mixture of flowers of sulphur and fine aluminium filings, containing two of sulphur to one of aluminium, is placed in a cylindrical glass tube 35mm. long. By means of a piston, a pressure of 20 kilogrammes per square centimetre (284lb. per square inch) was applied. It was only necessary to connect the column for 10sec. to the poles of a 25 cell battery, for the resistance originally infinite to be reduced to 4,000 ohms.

The arrangement shown in Fig. 56 illustrates another order of experiment. Two rods of copper were oxidised in the flame of a Bunsen burner, and were then arranged to lie across each other, as shown, and were connected to the terminals of the arm of a Wheatstone bridge, the high resistance of the circuit being due to the layers of oxide. Amongst the many measurements made, I found, in one case, a resistance of 80,000 ohms, which, after exposure to the influence of the electric spark, was reduced to 7 ohms. Analogous effects are obtained with oxidised steel rods. Another pretty experiment is to place a cylinder of copper, with an oxidised hemispherical head, on a sheet of oxidised copper. Before exposure to the influence of the electric spark, the oxide offers considerable resistance. The experiment can be repeated several times by

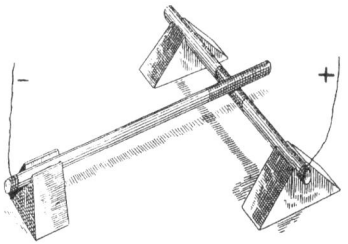

Fig. 56.

merely moving the cylinder from one place to another on the oxidised sheet of copper, thus showing that the phenomenon only takes place at the point of contact of the two layers of oxide.

In conclusion, it may be worth noting that, for most of the substances enumerated, an elevation of temperature diminishes the resistance, but the effect of a rise of temperature is transient, and is incomparably less than the effect due to currents of high potential. For a few substances the two effects are opposed.

A second article by Mr. Branly in *La Lumière Electrique* was abstracted in *The Electrician* for August 21, 1891, as follows :—

In a preceding article I showed that certain substances undergo an increase in conductivity under various electrical influences, and that these substances are numerous. The increase in conductivity varies with the energy of the exciting source. If the electric influence is due to the passage of a continuous current, the increase in conductivity is greater the greater the electromotive force of the battery employed. There is, however, no proportionality, the increased conductivity growing more rapidly than the number of cells, and tending quickly to a maximum. If the electric action consists in the passage of discharge currents in metallic rods, as in arrangement A (Fig. 53, p. 97), the conductivity increases with the length of spark at S, and it also increases when the rods are brought nearer the sensitive substance. Successive sparks are additive in their effects, although, if the action of the first one has been very powerful, the resistance is sometimes almost immediately reduced to a minimum.

Restoration of Original Resistance.—The conductivity causes by the various electrical influences lasts sometimes for a long period (24 hours or more), but it is always possible to make it rapidly disappear, particularly by a shock.

The majority of substances tested showed an increase of resistance on being shaken previous to being submitted to any special electrical influence, but after having been influenced the effect of shock is much more marked. The phenomenon is best seen with the metallic filings, but it can also be observed with metalised ebonite sheets with mixtures of liquid insulators and metallic powders, mixtures of metallic filings and insulating substances (compressed or not compressed), and finally with solid bodies.

I observed the return to original resistance in the following manner :—The sensitive substance was placed at K (Fig. 53), and formed part of a circuit which included a Daniell cell and galvanometer. At first no current passes. Sparks are then caused at S, and the needle of the galvanometer is permanently deflected. On smartly tapping the table supporting the ebonite cap in which the sensitive substance is contained, the original condition is completely restored. When the electric action has been of a powerful character, violent blows are necessary. I employed for the purpose of these shocks a hammer fixed on the table, the blows of which could be regulated.

With some substances, when feebly electrified, the return seemed to be spontaneous, although it was slower than the return of the galvanometer needle to equilibrium. This restoration of the original resistance is attributable to surrounding trepidations, as it was only necessary to walk about the room at a distance of a few metres, or to shake a distant wall. This spontaneous return to original resistance after weak electrical action was visible with a mixture of equal parts of fine selenium and tellurium powders. The restoration of resistance by shock was not observable so long as the electrical influence was at work.

After having been submitted to powerful electric action, shock does not seem to entirely restore substances to their original state, in fact, the substances generally show greater sensitiveness to electric action. Thus, a mixture of colza oil and antimony powder being exposed to the influence of arrangement A, a spark of 5mm. was at first necessary to break down the resistance, but after the conductivity had been made to disappear by means of blows, a spark of only 1mm. was sufficient to again render the substance conducting. Finely powdered aluminium has an extremely high resistance. A vertical column of powdered aluminium 5mm. long of 4 sq. cms. cross-section, submitted to considerable pressure, completely stopped the current from a Daniell cell. The influence of arrangement A produced no effect, but, by direct contact with a Leyden jar, the resistance was reduced to 50 ohms. The effect of shock was then tried, and after this the sparks produced by arrangement A were able to reduce the resistance.

The following experiment is also of the same kind. Aluminium filings placed in a parallelipidic trough completely stopped the current from a Daniell cell, and the resistance offered to a single cell remained infinite after the trough had been placed in the circuit of 25 sulphate of mercury cells for 10sec. The aluminium was next placed in circuit with a battery of 75 cells ; a single Daniell cell was then able to send a current through the substance. The original resistance was restored by shock, but not the original condition of things, since a single cell was able to send a current after the aluminium had been circuited for 10sec. with a battery of only 25 cells. I may add that if the restoration of resistance was brought about by a *violent* shock, it was necessary to place the aluminium in circuit with 75 cells for one minute before the resistance was again broken down.

It must be observed that electrical influence is not always necessary to restore conductivity after an apparent return to the original resistance, repeated feeble blows being sometimes successful in bringing this about. Both in the case of slow return by time and sudden return by shock, the original value of the resistance is often increased. Rods of Carré carbon, 1 metre long and 1mm. in diameter, were particularly noticeable for this phenomenon.

Return to Original Resistance by Temperature Elevation.—A plate of coppered ebonite rendered conducting by electricity, and placed close to a gas jet, quickly regained its original resistance. A solid rod of resin and aluminium, or of sulphur and aluminium, rendered conducting by connection to the poles of a small battery will regain its original resistance by shock; but if the conducting state has been caused by powerful means, such, for instance, as direct contact with a Leyden jar, shock no longer has any effect, at least such a shock as the fragile nature of the material can stand. A slight rise of temperature, however, has the desired result. By suitably regulating the electric action it is possible to get a substance into such a condition that the warmth of the fingers suffices to annul conductivity.

Influence of Surroundings.—Electric action gives rise to no alteration of resistance when the substance is entirely within a closed metal box. The sensitive substance, in circuit with a Daniell cell and a galvanometer, is placed inside a brass box (Fig. 54, p. 97). The absence of current is ascertained, the circuit broken, and the box closed. A Wimshurst machine is then worked a little way off, and will be found to have had no effect. The same result will be obtained if the circuit is kept closed during the time the Wimshurst machine is in operation. If a wire connected at some point to the circuit is passed out through a hole in the box to a distance of 20cm. to 50cm., the influence of the Wimshurst machine makes itself felt. On tapping the lid to restore resistance the galvanometer needle remains deflected so long as the sparks continue to pass. If, however, the wires are pushed in so that they only project a few millimetres, the sparks still passing, a few taps suffice to bring back the needle to zero. On touching the end of the wire with the fingers or a piece of metal conductivity is immediately restored. The movements of the galvanometer needle were rendered visible in these experiments by looking through a piece of wide mesh wire gauze with a telescope. The respective position of the things was also reversed ; that is to say, a Ruhmkorff coil and a periodically discharged Leyden jar were placed inside, and the sensitive substance outside, the box, with the same results.

In some later experiments with a larger metallic case (Fig. 57), and with the Daniel cell, sensitive substance, delicate galvanometer, and Wheatstone Bridge placed inside, I found

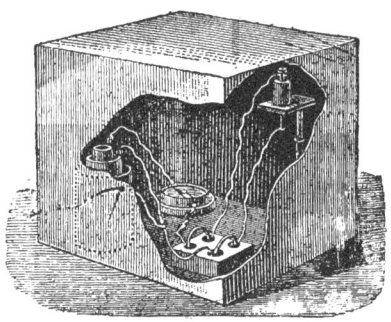

Fig. 57.

that a double casing was necessary in order to absolutely suppress all effects. A glass covering afforded no protection.

Considerations on the Mechanism of the Effects Produced.— What conclusions are we to draw from the experiments described ? The substances employed in these investigations were not conductors, since the metallic particles composing them were separated from each other in the midst of an insulating medium. It was not surprising that currents of high potential, and especially currents induced by discharges, should spark across the insulating intervals. But as the conductivity *persisted* afterwards, even for the weakest thermo-electric currents, there is some ground for supposing that the insulating medium is transformed by the passage of the current, and that certain actions, such as shock and rise of temperature, bring about a modification of this new state of the insulating body. Actual movement of the metallic particles cannot be imagined in experiments where the particles, in a layer a few millimetres thick, were fixed in an invariable relative position by extreme pressures, reaching at times to more than 100 kilogrammes per sq. cm. (1,420lb. to the square inch). Moreover, in the case of solid mixtures, in which the same variations of resistance were produced, displacement seems out of the question. To explain the persistence of the conductivity after the cessation of the electrical influence, are we to suppose in the case of metallic filings a partial volatilisation of the particles creating a conducting medium between the grains of metal? In the case of mixtures of metallic powders, and insulating substances agglomerated by fusion, are we to suppose that the thin insulating layers are pierced by the passage of very small sparks, and that the holes left behind are coated with conducting material? If this explication is admissible for induced currents, it must hold good for continuous currents. If so, we must conclude that these mechanical actions may be produced by batteries of only 10 to 20 volts electromotive force, and which only cause an insignificant current to pass. The following experiment is worth quoting in this connection :—

A circuit was formed by a Daniell cell, a sensitive galvanometer, and some aluminium filings in an ebonite cup, The galvanometer needle remained at zero. The filings were cut out of this circuit, and switched for one minute into circuit with a battery of 43 sulphate of mercury cells. On being replaced in the first circuit, the filings exhibited high conductivity. The result was the same when 10 or 20 cells were employed, or when the current was diminished by interposing in the circuit a column of distilled water, 40cm. long and 20mm. in diameter. The cells used (platinum, sulphate of mercury, sulphate of zinc, zinc) had a high internal resistance. Thus, 43 cells (60 volts), when short circuited, only gave a current of 5 milliamperes. The same battery, with the column of distilled water in circuit only, caused a deflection of 100mm. on a scale one metre off, with an astatic galvanometer wound with 50,000 turns. We can, therefore, see how infinitesimally small the initial current must have been when the filings were added to the circuit. The battery acted, therefore, essentially by virtue of its electromotive force.

If mechanical displacement of particles or transportation of conducting bodies seem inadmissible, it is probable that there is a modification of the insulator itself, the modification persisting for some time by virtue of a sort of " coercive force." An electric current of high potential, which would be completely stopped by a thick insulating sheet, may be supposed to gradually traverse the very thin dielectric layers between the conducting particles, the passage being effected very rapidly if the electric pressure is great, and more slowly if the pressure is less.

Increase of Resistance.—An increase of resistance was observed in these investigations less often than a diminution; nevertheless, a number of frequently repeated experiments enable me to say that increase of resistance is not exceptional, and that the conditions under which it takes place are well defined. Short columns of antimony or aluminium powder when subjected to a pressure of about 1 kilogramme per square centimetre (14·2lb. per square inch), and offering but a low resistance, exhibited an increase of resistance under the influence of a powerful electrification. Peroxide of lead, a fairly good conductor, always exhibited an increase ; so also did some kinds of platinised glass, while others showed alter-nate effects. For instance, a sheet of platinised glass, which offered a resistance of 700 ohms, became highly conducting after 150 sulphate of mercury cells had been applied to it for 10sec. This condition of conductivity was annulled by contact with a charged Leyden jar, and reappeared after again applying 150 cells for 10sec., and so on. Similar effects were obtained with a thin layer of a mixture of selenium and tellurium poured, when fused, into a groove in a sheet of mica placed between two copper plates. These alternations were always observed several times in succession, and at intervals of several days.

These augmentations and alternations are in no way incompatible with the hypothesis of a physical modification of the insulator by electrical influence.

APPENDIX III.

In connection with the branch of the subject dealt with on page 34, the following communications from Prof. Elihu Thomson and Dr. William J. Morton, M.D., which appeared respectively in the *Electrical Engineer*, of New York, July 4 and October 24, 1894, will be read with interest. Prof. Thomson writes :—

CURIOUS EFFECTS OF HERTZIAN WAVES.

In the issue of the London *Electrician* of June 8, 1894, under the heading, "Hertzian Waves at the Royal Institution," the following remark occurs : "It is wholly probable, as Dr. Lodge suggests, that Hertzian waves may often have manifested themselves in physical laboratories to the annoyance of the workers, &c."

I may mention in this connection that in 1877, if I remember the year correctly, while working a Ruhmkorff induction coil, one terminal of which was grounded and the other terminal of which was attached to an insulated metallic body, Prof. Houston and I noticed that when the sparks were passing between the terminals of the coil, it was possible not only to obtain minute sparks from all metallic bodies in the immediate neighbourhood, that is, in the same room, but that delicate sparks could be taken by holding in the hand a small piece of metal near metallic objects in many other rooms and on different floors in the building, although the pieces were not connected to ground. These could only have been Hertzian effects, but there was no recognition of their true character at the time, though the effects were seen to be connected with the very quick charging and discharging of the insulated body. An account of these experiments was, I think, published in the *Journal* of the Franklin Institute at the time.

I desire also to mention, as coming under my notice within the past year, a curious and rather amusing illustration of the principle upon which the beautiful instrument for detecting the presence of electric oscillations, devised by Dr. Lodge and called by him the " coherer," is based.

It was reported to me when in Philadelphia that a certain electro-plater had found that he could not pursue his silver plating operations during thunderstorms, and that if he left his plating over night and a thunderstorm came up the work was invariably ruined. I was disposed to be thoroughly sceptical, and expressed my disbelief in any such effect. Being urged, however, I went to the silver-plater's shop, which was a small one, and questioned the silver-plater himself concerning the circumstance which had been reported. While it was evident that he was not a man who had informed himself electrically, I could not doubt that, after conversing with him, he had indeed been stating what was perfectly true, namely, that when his operations of plating were going on and a thunderstorm arose, his batteries, which were Smee cells, acted as though they were short-circuited, and the deposit of metal was made at too rapid a rate. The secret came out on an inspection of his connections. The connections of his batteries to his baths were made through a number of bad contacts which could not fail to be of high resistance under ordinary conditions. I could readily see that virtually he was working through a considerable resistance and that he had an excess of battery power for the work. Under these circumstances a flash of lightning would cause coherence of his badly contacting surfaces, and would improve the conductivity so as to cause an excessive flow of current, give a too rapid deposit, and—as he put it—" make the batteries boil."

The incident suggests the use of Dr. Lodge's ingenious instrument in the study of the waves which are propagated during thunderstorms, of which waves we have practically little or no information.

APPENDIX III. *(continued)*.

Dr. Morton's communication is as follows :—

HERTZIAN WAVES, CARBON MICROPHONES AND "COHERERS."

About 18 months ago I put into use in my office the Vetter method of controlling the strength of the current derived from the Edison 110 volt system of electrical distribution. The controlling devices were a 16 c.p. lamp and a pulverised carbon rheostat. By these means a milliampere, or fraction thereof, up to 100 or more, if desirable, can be administered to a patient (*see* diagram, Fig. 58, on next page).

On several occasions when the electrodes of the system above described were permanently attached to some part of a patient's person and a spark was being administered to another patient seated upon a platform charged by an influence machine, some 15ft. distant in the same room, the first patient would exclaim and protest against receiving a considerable shock. On one occasion, when the continuous current electrode was in the neighbourhood of a patient's temple, the patient experienced the sensation of a flash of light; on other occasions muscular contractions were produced, always simultaneously with the spark. Also upon the occurrence of the spark and shock the needle of the milliamperemeter, a vertical one and calibrated to a wide range of movement over 5 milliamperes, flew across the scale from, for instance, 2 to 5 milliamperes and remained at the higher reading. That a spark occurring 15ft. away should cause a shock to a person in an independent circuit excited my wonder; it was inexplicable and yet so certain to occur that I was obliged to abandon the use of the two pieces of apparatus at the same time.

At last, when time permitted, I set out to investigate. I sought for an ordinary induction circuit of parallel wiring and found none. I then suspected the microphonic rheostat of pulverised carbon and having cut it out of circuit I substituted for it a water rheostat. The phenomena now failed to occur. Replacing the carbon rheostat and putting a telephone in circuit I adjusted the milliammeter to read 2 milliamperes, causing an assistant to evoke the distant spark. All was now clear. At each spark the needle jumped forward and a distinct telephone click was heard from the telephone receiver. I observed that the first jump of the needle was the longest as well as the first click in the receiver

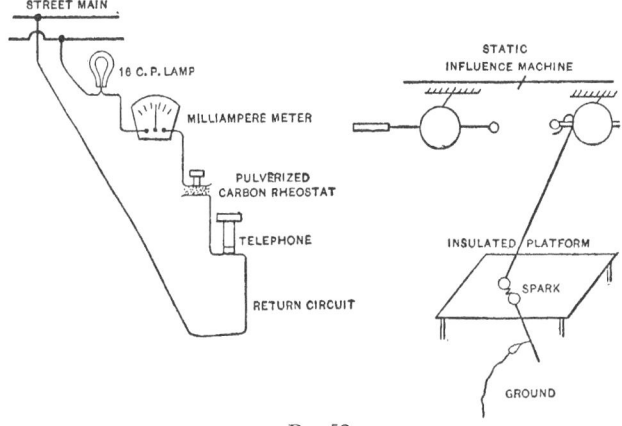

STREET MAIN

16 C. P. LAMP

MILLIAMPERE METER

PULVERIZED
CARBON RHEOSTAT

TELEPHONE

RETURN CIRCUIT

STATIC
INFLUENCE MACHINE

INSULATED PLATFORM

SPARK

GROUND

FIG. 58.

the loudest, both needle jump and click, dying away gradually at each successive spark until they ceased at from the twentieth to thirtieth. To turn the rheostat off and then on again rendered the experiment repeatable. The reading of the meter, best adapted to success, was about 5 milliamperes though 20 to 50 yielded good results.

Unable to furnish any reason why the electric radiation of a distant spark should reduce the resistance of pulverised carbon I refrained from publishing the bare observation in the hopes of finding an explanation by further experimentation, merely noting to friends the delicacy of the pulverised carbon rheostat as a detector of Hertzian waves and making some further experiments with it and a telephone receiver in circuit in this direction.

The recent publication of the brilliant researches of Dr. Oliver J. Lodge now makes the entire matter clear. Dr. Lodge describes a new form of microphonic detector of Hertzian waves, consisting of two or more pieces of fairly clean metal in light contact and connected to a voltaic cell, a film of oxide of the metal intervening between the surfaces, ' so that only an insignificant current is allowed to pass."

He writes: " Now let the slightest surging occur, say, by reason of a sphere being charged and discharged at a distance of 40 yards; the film at once breaks down—perhaps not completely, that is a question of intensity—but permanently."

This detector, Dr. Lodge terms a " coherer " because of the partial metallic cohesion above described. Upon this point he says: " A bad contact was at one time regarded as a simple nuisance." . . . "Hughes observed its sensitiveness to sound waves, and it became the microphone. Now it turns out to be sensitive to electric waves, if it be made of any oxidisable metal (not of carbon) and we have an instrument which might be called a micro-something but which, as it appears to act by cohesion, I call at present a coherer." The cohesive result between the metallic surfaces is also referred to as a "welding effect of an electric jerk." In the volume just published, entitled " The Work of Hertz and Some of His Successors," reprinted from *The Electrician*, London, this foot note is added on p. 30: " FitzGerald tells me that he has succeeded with carbon also."

My experiments would seem to fully demonstrate that carbon as well as metals may act as coherers. At some recent trials the editors of the *Electrical Engineer* were present and were fully satisfied as to the swinging up of the needle of the milliamperemeter and the click in the telephone receiver, by repeated tests.

The experimental side of the subject has been so exhaustively and admirably presented by Dr. Lodge (detailed in the publications referred to) that what is here said has no more than a secondary interest. But it may not prove amiss to gather together all the evidence which tends to demonstrate the influence of disruptive discharges upon neighbouring bad contacts conveying currents. As Lodge points out, fuses may easily be "blown out" in this manner. This has occurred to me on a number of occasions with 10 ampere fuses. Under proper conditions of sparking surfaces and circuit a short spark might suffice.

May it not also be the fact that the fuses melted during thunderstorms in their neighbourhood are melted by reason of the effect of the electric radiations or surgings of the lightning stroke throwing a rush of the current already in the circuit through the fuse rather than by the addition of any new current to the circuit by the atmospheric electricity itself. In this connection Lodge writes: "There are some who think that lightning flashes can do none of these secondary things. They are mistaken." In this as in other directions the new facts have a practical bearing and a pursuit of further experiments may lead, as often happens, to unexpected developments.

So far as carbon contacts are concerned and the fact that Hertzian waves, like mechanical motion, reduce their resistance, a curious problem is suggested as concerns the principle upon which some carbon transmitters act. An exclusive monopoly of all carbon transmitters is based upon the claim that the variations in resistance are produced by variations in *pressure* due to a mechanical force, viz., sound waves. If my experiments, above detailed, are exact, two facts appear:

1. That another form of motion, ether vibration, causes a variation of resistance of carbon contacts.

2. That it remains to be proved that variation of *pressure* is the only means of varying the current strength, for variation of molecular contact occurs in the present instance without any evidence that it is due to variation of pressure.

APPENDIX IV.

ON THE DISELECTRIFICATION OF METALS AND OTHER BODIES BY LIGHT.

Referring to a footnote to my Royal Institution lecture, on page 11, Messrs. Elster and Geitel have been good enough to call my attention to a great deal of work done by them in the same direction. To make amends for my ignorance of this work at the time of my Royal Institution lecture, and to make it better known in this country, I make abstract of their Papers as follows :—

Wiedemann's Annalen, 38, p. 40.—"*On the Dissipation of Negative Electricity by Sun- and Daylight.*"

With a view to Arrhenius' theory concerning atmospheric electricity, we arranged experiments on the photo-electric power of sunlight and diffuse daylight at Wolfenbüttel from the middle of May to the middle of June, 1889. Hoor alone had observed the effect of sunlight ; other experimenters had failed to find it, but we find a discharging effect even in diffuse daylight.

We take an insulated zinc dish, 20 cm. diameter, connect it to a quadrant electrometer or an Exner's electroscope, and expose it in the open so that it can be darkened or illuminated at pleasure. Sunlight makes it lose a negative charge of 300 volts in about 60 seconds. A positive charge of 300 volts is retained. The dissipation of negative electricity ceases in the dark, and is much weakened by the interposition of glass. But light from the blue sky has a distinct effect. Fill the dish with water, or stretch a damp cloth over it, and the action stops. A freshly-scrubbed plate acquires a positive charge of $2\frac{1}{2}$ volts, which can be increased by blowing.

With freshly-cleansed wires of zinc, aluminium, or magnesium attached to the knob of the electroscope, a permanent negative charge is impossible in open sunlight.   Indeed, magnesium shows a dissipating action in diffuse evening light. Such wires act like glowing bodies.   Exposing an electroscope so provided in an open space it acquires a positive charge from the atmosphere.   No abnormal dissipation of positive electricity has been observed.

Wied. Ann., 38, p. 497.—Continuation of Same Subject.

Our success last time was largely due to the great clearness of the sky in June, and we wished to see if we could get the same effect at the beginning of the winter.

The following is our summary of results :

Bright fresh surfaces of the metals zinc, aluminium, magnesium were discharged by both sun- and daylight when they were negatively charged ;   and they spontaneously acquired a positive charge, whose amount could be increased by blowing.*   A still more notable sensitiveness to light is shown by the amalgams of certain metals, viz., in the order of their sensitiveness, K, Na, Zn, Sn.   Since pure mercury shows no effect, the hypothesis is permissible that the active agent is the metal dissolved in the mercury.   If so, the following are the most active metals :—

K, Na (Mg, Al), Zn, Sn.

All other metals tried, such as Sn, Cd, Pb, Cu, Fe, Hg, Pt, and gas carbon, show no action.   The same is true of nearly all non-metallic bodies ; but one of them—namely, the powder of *Balmain's luminous paint*—acted remarkably well in sunlight. Of liquids, hot and cold water, and hot and cold salt solution were completely inactive ; consequently wetting the surface of metals destroys their sensibility to light.

The illumination experiments can be arranged in either of two ways.   For experiments in free space we use zinc, aluminium, or magnesium wires, or small amalgamated spheres of zinc provided with an iron rod.   With these it can be easily shown that the illuminated surface of certain metals act in the same way as a flame collector.

---

* A fact noticed by Bichat and Blondlot.

For demonstration experiments the apparatus described*
is better, and with this we show the following :—

Amalgamated zinc, negatively charged, discharges almost
instantly in sunlight; and if near a positively-electrified body
charges itself positively.

The same thing happens, though more slowly, in diffuse
daylight. Red glass stops the action, but the following let
some through :—Selenite, mica, window glass, blue (cobalt)
glass.

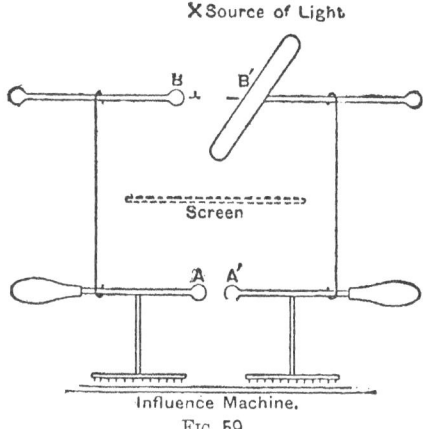

X Source of Light

Screen

Influence Machine.

FIG. 59.

Explanation of Fig. 59.—B' is a brightly polished amalgamated zinc
plate attached to the negative pole of a Holtz machine, with the positive
knob from 6 to 10 centimetres distant. The source of a light is a strip of
burning magnesium ribbon 30 to 50 centimetres away. Whenever the
spark is able just to choose the path B B', light shining on the zinc plate
checks it and transfers the spark to A A'.

*Wied. Ann., 39, p. 332.—On a Checking Action of Illumination
on Electric Spark and Brush Discharge.*

If sparks are just able to occur between a brass knob and a
clean amalgamated zinc cathode, illumination of the latter by

* In this apparatus the mercury amalgams of K and Na are run through
a fine funnel, so that the freshly-formed surface of the drops may be
illuminated. Under these circumstances, while pure mercury fell from
−185 to −175 volts in 30sec., amalgam of zinc fell from −195 to −116
in 15sec., amalgam of sodium fell from −195 to 0 in 10sec., and an
amalgam of potassium fell from −195 to 0 in 5sec.

ultra-violet light tends to check them. [This is a curious inversion of Hertz's fundamental experiment on the subject. It is an effect I have not yet observed ; but Elster and Geitel's arrangement differs from mine* in that the surfaces are at a steady high potential before the spark, so that light can exert its discharging influence, whereas in mine the surfaces were at zero potential until the spark-rush occurred.   Hertz's arrangement was more like mine, inasmuch as he illuminated the knobs of an induction coil on 'the verge of sparking.   It appears, then, that whereas the action of light in discharging negative electricity from clean oxidisable metallic surfaces is definite enough, its influence on a spark discharge differs according to the conditions of that discharge—in cases of " steady strain " it tends to hinder the spark; in cases of " sudden rush " it tends to assist it.—O. J. L.]

Wied Ann., 41, p. 161.—On the use of Sodium-Amalgam in Photo-electric Experiments.

Elster and Geitel have repeated some of Righi's experiments on the discharge of negative electricity from metals in rarefied air, and find, in agreement with him, that a reduction of pressure to about one millimetre increases the discharge velocity about six or seven times.   They proceed to try sodium-amalgam exposed to daylight in exhausted tubes, and describe apparatus for the purpose.   Such an arrangement simply cannot hold a negative charge in bright daylight, even although it be unprovided with quartz windows.   Even paraffin lamps and sodium flames exert some action.

They observe that under the action of light the boundary surface of the metal and glass changes, and the metal begins to cling to the glass.   They suppose that Warburg's vacuum tubes of pure sodium may behave similarly, and show photo-electric sensibility.

The Same, p. 166.—On a Checking Action of Magnetism on Photo-Electric Discharge in Rarefied Gases.

The authors point out analogies between the above effects and those they had observed in the action of glowing bodies in air, and they mention Lenard and Wolf's experiments

---

* *See* Fig. 7, page 9

Wied. Ann. XXXVII., p. 443), tending to show that the effect is due to a disintegrating or evaporative effect of light on surfaces. Elster and Geitel had observed that the discharging power of glowing bodies was diminished by application of a magnetic field, the effect being the same as if the temperature was lowered ; and they proceed to try if the discharge of negative electricity from illuminated surfaces in highly-rarefied gas could also be checked or hindered by a magnetic field. They find that it can.

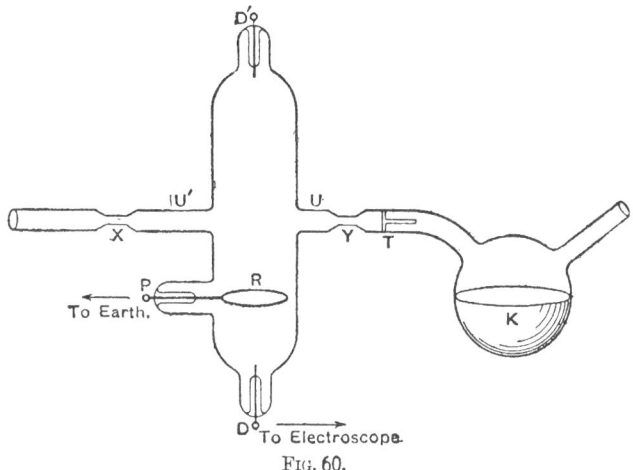

Fig. 60.

Explanation of Fig. 60.—The sodium and mercury are introduced through the tube S into the globe K. The tube S is then closed, a pump applied to X, and exhaustion carried on for some days. T is an open funnel sealed into the tube (as is done in some vacuum tubes made by Holtz) to show a curious unilateral conductivity of rarefied gas. The object of this funnel is to permit metal from the interior, free from scum, to be introduced from K to D when the whole is tilted. Thus a bright surface is exposed to the earth ring R. It can be charged negatively, and its leak under illumination be measured, through the terminal D. Sometimes the tube is inverted, so that the active surface may be at D', further from the earth wire.

Using the light from sparks admitted through a quartz window into the vacuum tube when a negatively-charged amalgamated zinc surface was exposed near an earth-connected platinum ring, and between the poles of a small electromagnet, they found that when the tube was full of air at

10mm. pressure the magnet had but little effect, but that at 0·15mm., whereas without the magnet the charge of −270 volts disappeared completely in five seconds, when the magnet was excited it only fell about half that amount in the same time. With hydrogen at 0·24mm. the result was much the same, and at either greater or less pressure in both cases the magnet had less effect. In oxygen the loss of charge was not quite so rapid; and, again, at a pressure of 0·1mm., the magnet more than halved the rate. But in CO_2 the rapidity of loss was extreme,* Either at 1·1mm. or at 0·005mm. the charge of

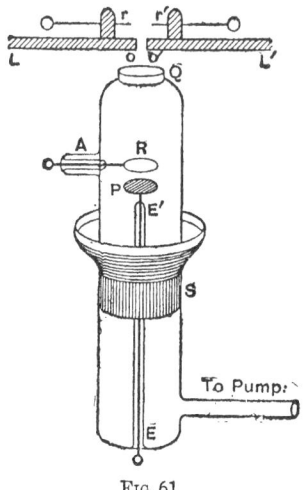

FIG. 61.

Explanation of Fig. 61.—P is the plate of amalgamated zinc, and **R** is the earth ring, as before. Ultra-violet light is introduced through a quartz window Q from a spark gap *r*. The vessel has a joint at the middle, so that the sensitive plate can be got at and changed. Magnet poles are applied outside this vessel in various positions.

270 volts leaked away completely in two seconds when the magnet was not excited; but in the latter case (low pressure) exciting the magnet reduced the speed by about one-half. At the pressure of 1·1mm. the magnet did not seem to produce an effect. With daylight the results are similar.

* Corresponding to the activity of this gas as found by Wiedemann and Ebert (*Wied. Ann.*, XXXIII., p. 258), in their researches on the influence of light on ease of sparking.

The authors then discuss the meaning of the result, and its bearing on the opposition hypotheses of Lenard and Wolf and of Righi. Lenard and Wolf's view is that the loss of negative electricity is due to dust disintegrated from the surface by the action of light, but whose existence they consider is established by an observed effect on steam jets. Righi, on the other hand, believes that gas molecules themselves act the part of electric carriers. Elster and Geitel consider that the magnetic effect observed by them supports this latter view, it being known that a magnet acts on currents through gases; and they surmise that the impact of light vibrations may directly assist electric interchange between a gas molecule and the surface, by setting up in them syntonic stationary vibrations, something like resonant Leyden jars. It is to be remembered that phosphorescent substances, such as Balmain's paint powder, exhibit marked photo-electric effect in daylight.

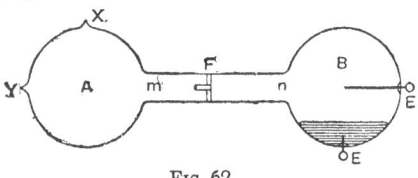

Fig. 62.

Explanation of Fig. 62.—A simpler arrangement, like the one above (Fig. 61), whereby clean liquid alkali metals can be introduced into the experimental chamber B, from the preliminary chamber A, through a cleansing funnel, F, which dips its beak into the interior.

The unilateral character of the electric motion, and the charging of neutral surfaces by light, require special hypotheses, concerning an E.M.F. at the boundary of gases and conductors, such as Schuster and Lehmann have made.

Weid. Ann. 42, p. 564.—Note on a New Form of Apparatus for Demonstrating the Photo-electric Discharging Action of Daylight.

A vacuum tube suitable for experiments with sodium amalgam or pure sodium, or the liquid sodium-potassium alloy, is described, with the aid of which a current (shown by the charge of an electroscope) can be maintained by a dry pile through the rarefied gas above the metal when it is illuminated from ordinary windows.

Wied. Ann. 43, p. 225.—On the Dependence of the Discharging Action of Light on the Nature of the Illuminated Surface.
Experiments also on differently-coloured lights. Summary of results. The photo-electrically active metals arrange themselves in the following order—Pure K, alloys of K and Na, pure Na. Amalgams of Rb, K, Na, Li, Mg (Tl, Zn); the same as their voltaic order. With the most sensitive term of the series a candle six metres off can be detected, and the region of spectral red is not inactive. The later terms of the series demand smaller waves, and even for potassium blue light gives a much greater effect than red. No discharge of positive electricity is observable with these substances.

Wied. Ann. 44, p. 722.—On the Dissipation of Electric Charge from Mineral Surfaces by Sunlight.
Hitherto only Balmain's paint powder has been observed to be active among non-metallic substances. Now they try other phosphorescent bodies, and arrive at the following results:—

Fluorspar is conspicuously photo-electric, both in sunlight and daylight, especially the variety of fluorite called *stinkfluss*.

Freshly-broken surfaces discharge much more rapidly than old surfaces.

Blue waves, and not alone the ultra-violet, have a perceptible effect on fluorspar.

In a vacuum the mineral loses its photo-electric sensibility and its conductivity too. Contact with damp air restores its sensibility. Moistening with water weakens but does not destroy the sensitiveness. On the other hand, igniting the mineral destroys both its photo-electric power and its exceptional phosphorescent property.

Distinct traces of photo-electric power are shown by the following minerals also: Cryolite, heavy spar, celestine, arragonite, strontianite, calcspar, felspar and granite.

The hypothesis that the power of phosphorescing when illuminated is approximately a measure of the discharging power of light has been verified in many cases; the exceptions can probably be explained by the influence which the electrical conductivity of the illuminated substance exerts on the rate of discharge of electricity from its surface. This agreement confirms the view expressed by us on the occasion of experiments with Balmain's paint, that, during electrical discharge by light, actions take place which are analogous to those of resonance. Messrs. Wiedemann and Ebert had previously been led by other considerations to the same conclusion.

We are compelled by the results of the present experiments to conclude that a more rapid discharge of electricity into the atmosphere takes place in sunlight than in darkness from the surfaces of the earth, which is composed of mineral particles charged, as the positive sign of the slope of atmospheric potential indicates, with negative electricity.

It seems to us evident that there exists a direct electric action of sunlight upon the earth, and that we have given experimental evidence in favour of the theory put forward by von Bezold and Arrhenius, according to which the sun acts on the earth, not by electrostatic or electro-dynamic

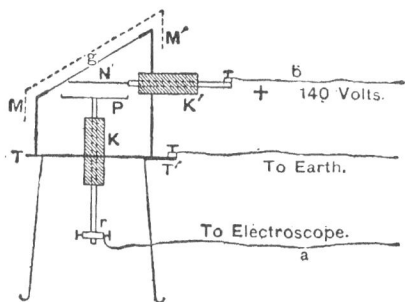

FIG. 63.

Explanation of Fig. 63.—Arrangement used by Elster and Geitel for exposing various phosphorescent minerals to daylight, while under inductive charge. They were put in powder in the tray P, and the transparent wiregauze N above them was charged positively from a battery. The metal cover MM' could be removed and replaced at pleasure, and the effect on a delicate quadrant electrometer connected to P observed. By this method considerable tension can be got up on the mineral surface, notwithstanding that it is close upon zero potential. The light effect depends on tension, not potential.

action-at-a-distance, which would involve difficulties of a theoretical character, but through the medium of the electrical forces of light waves. We hope soon to establish the consequences of this theory in meteorology in another Paper, giving the results of two years' observations on the intensity of the most refrangible rays of sunlight and of the slope of atmospheric potential.

Wied. Ann., 48, p. 338.—Experiments on the Gradient of Atmospheric Potential and on Ultra-Violet Solar Radiation.

Elster and Geitel describe the observations they have made for two years on solar radiation, at observing stations of low and high altitude, as tested by its electrical discharging power ; and they plot curves of such effective radiation for days and months along with the curves of atmotpheric potential observed at the same places. These curves are of much interest, and need study. Incidentally they find that, of the whole effective solar radiation, 60 per cent. was absorbed at altitudes above 3,100 metres ; 23 per cent. of the remainder was absorbed in the layer between this and a station at 1,600 metres ; and 47 per cent. was absorbed between this and 80 metres above sea level. Or, in other words, of 236 parts which enter the atmosphere 94 reach the highest observing station (Sonnblickgipfel), 72 the middle one (Kolm-Saigurn), and 38 the lowest (Wolfenbüttel). They discuss the question as to how far the daily variation of terrestrial magnetism is due to electrical currents in the atmosphere excited by sunshine and other meteorological matters.

[The Paper and plates are worthy of reproduction in full in the *Philosophical Magazine.*]

Weid. Ann., 46, p. 281.—On the Behaviour of Alkali Metal Cathodes in Geissler Tubes; On Photo-Electric Discharge in a Magnetic Field ; and On the Measure of Photo-Electric Currents in Potassium Cells by means of a Galvanometer.

Results :—The resistance of a Geissler tube provided with a cathode surface of pure alkali metal is diminished by the light from the sparks of an induction coil ; especially when the pressure is ·1 to ·01 mm. of mercury. The resistance which rarefied gas opposes to an electric current in a magnetic field is greatest in the direction normal to the magnetic lines. The changes of resistance effected by any kind of light in a vacuum tube with alkali metal cathode can be measured galvanometrically. (A Daniell cell gives 100 divisions on a Rosenthal galvanometer when coupled up through such an illuminated tube, each division meaning about 10^{-10} ampere.)

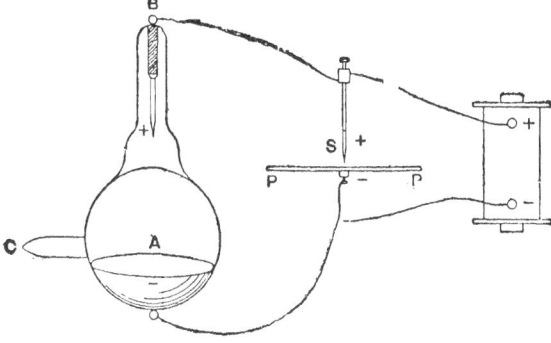

Fig. 64.

Explanation of Fig.64.—A vacuum tube of rarefied hydrogen containing alkali metal as cathode, say the liquid K—Na alloy, or solid K or Na. A spark gap at S serves as alternative path, and a stream of sparks can occur to the plate P in the dark.   But when light falls on the surface A this stream of sparks can cease, showing that the resistance of the vacuum tube is diminished.

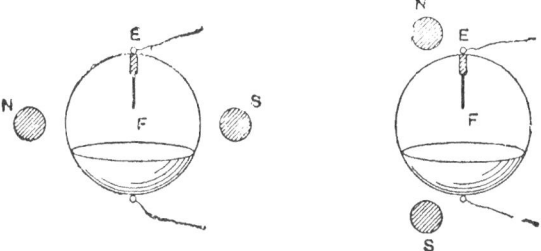

Fig. 65.

Explanation of Fig. 65.—Showing position of magnetic poles with respect to the vacuum tube discharge.   With the poles *across* the line of discharge, as in Fig. on left, excitation of the magnet opposes the leak from the surface   With the poles as in Fig. on right, the discharge is not much affected, it is even sometimes slightly increased.

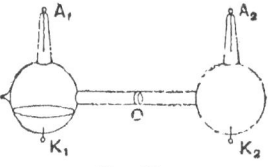

Fig. 66.

Explanation of Fig. 66.—Potassium vacuum bulbs containing $\frac{1}{3}$ millimetre of hydrogen mounted and connected to battery and galvanometer, and arranged as a photo-electric photometer.

Wied. Ann., 48, p. 625. On the Photo-Electric Comparison of Sources of Light.

Attempts to make such a potassium cell into a photometer.

Wied. Ann. 52, p. 433. Further Photo-Electric Experiments.

Plates of platinum, silver, copper need exceedingly ultra-violet light before they show any photo-electric power ; zinc, aluminium, magnesium show it for visible violet and blue light ; the alkali metals, in an atmosphere of rarefied hydrogen, advance their range of sensibility into the spectral red ; while under the most favourable conditions they show a sensibility only inferior to that of the eye itself. The authors now use galvanometric methods of measuring the effect, instead of only electrometers, and they arrive at the following results :—

(1) The three alkali metals Na, K, Rb, have different sensibility for differently-coloured lights. For long waves their order of sensibility is Rb, Na, K ; though rhubidium is far exceeded by the other two metals in white light.

(2) Illumination of a plane alkali-metal cathode surface with polarised light causes greatest discharge if the plane of polarisation is normal to plane of incidence ; and least, if the two coincide.

[This is a most remarkable observation. Its probable meaning is that the electric oscillations of light are photo-electrically effective in so far as they are normal to the surface on which they act ; while electric oscillations tangential to the surface are scarcely operative. Different angles of incidence must be tried before the proof is complete.—O. J. L.]

(3) Electric oscillations of very short period, such as are given by a Hertz oscillator, are commutated by illumination in the presence of alkali metals in rarefied gas, so as to be able to set up a constant electric tension in the gas.

[A Zehnder tube* was used, and the momentary phases of the oscillation during which the metal is negatively charged are apparently taken advantage of by the illumination.]

(4) The photo-electric dissipation showed by powdered fluorspar is dependent on the colour of the mineral, in such a way that the deepest blue, violet or green specimens are the most sensitive.

---

* See Fig. 13, p. 15.

APPENDIX V.

PHOTO-ELECTRIC RESEARCHES OF M. AUGUSTE RIGHI.*

M. Righi has observed the following facts : (1) That ultra-violet rays reduce to sensibly the same potential two metals placed near each other (plate and gauze parallel and close) ; (2) That several photo-electric couples of this kind can form a battery : (3) That a simple metallic plate charges itself positively under the influence of radiation ; (4) That a voltaic arc formed with a zinc rod gives the strongest effect, while the sun gives none.

Besides these facts he finds :—(a) That certain gases and vapours, such as coal-gas and CS_2, absorb the active rays strongly.

(b) That if the discharging body is easily movable it recedes like an electric windmill.

(c) A film of gypsum interposed between gauze and plate charges itself negatively on the side facing the negatively-charged plate.

(d) Radiation produces its discharging effect even on non-conductors (ebonite and sulphur). With glass, resin and varnishes the action is feeble, or nearly nothing.

(e) If the experiment is made with a copper gauze and a zinc plate, the phenomenon nearly disappears on varnishing the gauze. His hypothesis is that radiation produces convection of negative electricity, the carriers being molecules of air.

(f) The carrying molecules move along the lines of force, and throw electric shadows. To show this he varnishes a zinc cylinder, all except a generating line, charges it negatively to 1,000 volts with a dry pile, and places it parellel to a large

---

earth-connected plane, which has a narrow rectangular portion insulated from the rest and communicating with an electrometer. Light only acts on the uncovered line of the cylinder, and on turning the cylinder round the electrometer is only deflected when it is exposed to some of the (circular) lines of force emanating from the active line of the cylinder.

(*g*) Radiation charges positively an insulated metal, even when it is an enclosure with walls of the same metal; the metal being certainly uncharged at the beginning of the experiment. The same occurs with sulphur and ebonite. If there is a feeble initial *plus* charge, radiation increases it.

(*h*) While the discharging power of radiation for negative electricity is strongest with zinc and aluminium, and slower with copper and gold, following the Volta series; the E.M.F. set up by radiation, when it charges things positively, is greatest with gold and carbon, and less with zinc and aluminium; again following the Volta series, but inversely.

(*i*) If radiation falls on an insulated metal plate connected with an electrometer, in an enclosure of the same metal, the positive electrification shown by the deflection of the electrometer is greater as the plate is further from the walls of the enclosure. The action stops when the metal has attained a certain electric density, constant for a given metal; so the potential of a plate is natural.y higher as its capacity is less. It is thus established that radiation acts on the particles of gas in contact with a conductor; they go away with a negative charge, leaving *plus* on the conductor, until an electric density sufficient to balance this action is attained.

(*j*) It is probable that if the solar rays do not produce an effect it is because of the absorbing action of the atmosphere. In fact, if one places a tube whose ends are glazed with selenite between the source of light and the metals being experimented on, the effects become sensibly stronger when the tube is exhausted.

APPENDIX VI.

ELLIPTICALLY POLARISED ELECTRIC RADIATION.

Since the delivery of my lecture to the Royal Institution, on June 1st, Herr Zehnder has published* a mode of getting elliptically and circularly polarised electric radiation. He takes a couple of plane polarising grids, such as are depicted in Fig. 21, page 37, and places them parallel to each other at a little distance apart with their wires crossed.

If the two grids are close together they will act like wire-gauze, reflecting any kind of polarised radiation equally; but if the warp and woof are an eighth-wave length apart, and the plane of the incident radiation is at 45° to the wires, the reflected radiation will be circularly polarised. A change in the circumstances will, of course, make it elliptical. Such a pair of grids acts, in fact, like a Babinet's Compensator.

---

* *Berichte der Naturforschenden Gesellschaft zu Freiburg i. B.*, Bd. IX. Heft 2, June 21, 1894.

K

APPENDIX VII.

ON MAGNETISATION PRODUCED BY HERTZIAN CURRENTS; A MAGNETIC DIELECTRIC:*

BY M. BIRKELAND.

" Two years ago† it was proved by conclusive experiments that Hertzian waves travelling along an iron wire magnetise transversely the very thin layer into which the alternating current penetrates, and whose thickness does not exceed some thousandths of a millimetre. Once proved that alternate magnetisation can be produced with such rapidity, other questions present themselves. One asks, for instance, if it is not possible to demonstrate in magnetic cylinders stationary magnetic waves analogous to the electric stationary waves along metallic wires."

The author finds that the conductivity of massive iron makes it an unsuitable substance, and uses instead a mixture of iron filings, or of chemically-obtained iron powder, with paraffin, to which he sometimes adds powdered quartz. This he moulds into cylinders, and inserts as the core of a spiral in an otherwise ordinary Hertz resonator.

Fig. 67 shows emitter and receiver drawn to scale; the magnetic cores are introduced into the spiral A, and their effect on the length of the resonator spark is observed. With this arrangement of exciter the *electric* effect of the spiral is negligeable, since it is well removed from electrostatic disturbance, and subject only to magnetic. The spiral is of 12

---

* Abstracted from *Comptes Rendus,* June 11, 1894, and communicated by Dr. Oliver Lodge.

† Why two years ago ? It was practically proved by Savart early in the century, and has been observed over and over again since. However, it is true that experiments have been more numerous and conclusive of late, and have been pushed to very high frequencies.—O. J. L.

well-insulated turns, the spark-gap is a micrometer with point and knob, and a pair of adjustable plates to vary the capacity for purposes of tuning.

He employed 12 different types of cylinder, all about 20 centimetres long, and 4 centimetres diameter.

1. A massive cylinder of soft iron.

2. A bundle of fine iron wires embedded in paraffin.

3—9. Six cylinders of the agglomerate of chemically-reduced iron in powder and paraffin, containing respectively 5, 10, 15, 20, 25 and 50 per cent. of iron.

Then for control experiments :—

10. A cylinder of agglomerate of zinc powder in paraffin, with 40 per cent. of zinc.

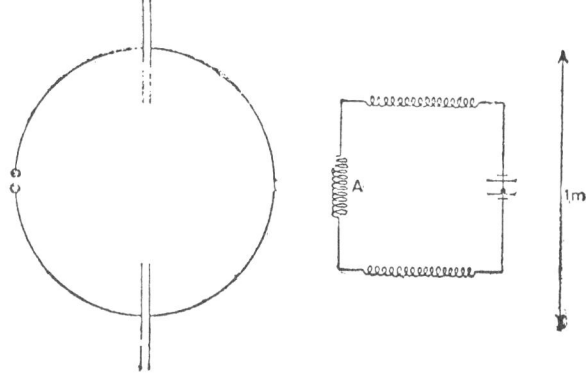

Fig. 67.

11. A cylinder of brass filings in paraffin, 20 per cent of metal.

12. A tube of glass, 4·5 centimetres diameter, filled with various electrolytes.

The manner of observing was as follows (the experiments were done in the laboratory of Hertz) :—

The resonator, with its spiral empty, was syntonised with the exciter, and the maximum spark measured. It was between 4 and 9 millimetres long in these experiments. Then one or other of the above cylinders was introduced and the spark length measured afresh.

Cylinder 1 did not affect the maximum spark-length.

Cylinders 2—4 reduced the maximum spark to $\frac{1}{10}$th of its former value ; 7 and 8 to $\frac{1}{100}$th, and No. 9 to $\frac{1}{200}$th of its former value (viz., from 9 millimetres to ·05 millimetre). Nos. 10 and 11 had but a feeble action, and reduced the spark from 8 to 7 millimetres.

Tube No. 12, filled with distilled water, scarcely affected the spark length ; the period of the secondary increases a little, but the maximum spark is the same as before, once syntony is re-established. Filled, however, with dilute sulphuric acid, containing 10, 20, or 30 per cent., the tube reduced the spark considerably, in each case about the same, viz., from 9 to 1·3 about. (Currents induced by Maxwellian radiation in electrolytes had been already observed by J. J. Thomson.)

While trying to re-establish syntony between primary and secondary, I found that the period of the resonator was considerably increased by the cylinders 2—4, but that the maximum spark length was much diminished. With the cylinders Nos. 5—9 in the spiral, it was no longer possible to establish syntony, "a fact which is certainly due to their considerable absorption of energy. Take, for example, cylinder 9 : electro-magnetic energy must converge rapidly towards it in order to be transformed, and the space finds itself empty of energy as air is exhausted of vapour in presence of an absorbing substance.

"This absorption is probably due to hysteresis in the ferruginous cylinders ; the development of Joulian heat, so typically shown by cylinder 12, being undoubtedly of the same order in cylinders 3—9 as in Nos. 10, 11.

"It is probably by reason of this absorption that I have not succeeded in establishing stationary magnetic waves in a circuit of ferro-paraffin."

If one of the cylinders 2—9, is wrapped in tinned paper before introducing it into the spiral A, its action is completely stopped. (These conducting cores *diminish* the period of the resonator ; it is much as if the spiral A were partially shunted out ; but the maximum spark returns as soon as syntony is reestablished.) To examine this further he enclosed the cylinder in drums of cardboard having fine wires either along generating lines, or along circular parallels. The latter suspended the action of an interior ferruginous cylinder, the former did not.

To find to what depths the magnetism penetrated, Birkeland inserted hollow ferruginous drums into A, measured their effect, and then plunged solid cylinders into them to see whether the effect increased.

He thus found that the magnetisation easily traversed 7 millimetres thickness of the 10 per cent. ferro-paraffin, and 5 millimetres of the 25 per cent.

The substance is comparable to a dieletric on the theory of Poisson-Mossotti.

" The results obtained with our magnetic dielectric invite to new researches "—such as the mechanical force excited by electric waves on a delicately-suspended ferro-paraffin needle, and the rate of propagation of Maxwellian waves through such a substance.

PHYSICS IN INDUSTRY

VOLUME III

THE INSTITUTE OF PHYSICS

¶ The object of the Institute is on the one hand to secure the recognition of the professional status of the physicist and to urge the importance of physics in industry, and on the other hand to co-ordinate the work of all the societies interested in physical science or its applications.

¶ The Institute grants Diplomas to Corporate Members in order to secure authoritative recognition of them as physicists engaged in professional work. Corporate Members are of two classes, Associates and Fellows, depending upon academic qualifications and professional experience. The Institute also registers students of physics.

¶ The Institute has compiled a Panel of Consulting Physicists and an Appointment Register.

¶ The Institute produces the *Journal of Scientific Instruments*, a monthly journal dealing with their principles, construction, and use, published by the Cambridge University Press.

¶ Regulations relating to admission may be obtained on application to the Secretary, 90 Great Russell Street, London, W.C. 1.

ELECTRICAL PRECIPITATION

A LECTURE DELIVERED BEFORE
THE INSTITUTE OF PHYSICS

BY

Sir OLIVER LODGE, D.Sc., F.R.S.

PHYSICS IN INDUSTRY

VOLUME III

OXFORD UNIVERSITY PRESS
LONDON: HUMPHREY MILFORD
1925

HUMPHREY MILFORD
OXFORD UNIVERSITY PRESS
London   Edinburgh   Glasgow   Copenhagen
New York   Toronto   Melbourne   Cape Town
Bombay   Calcutta   Madras   Shanghai

PRINTED IN GREAT BRITAIN BY MORRISON AND GIBB LTD., EDINBURGH.

INTRODUCTION

THE subject of Electrical Precipitation divides itself into two parts, the natural and the artificial. The natural has gone on in the atmosphere from time immemorial, small drops coalescing into larger ones, and all falling at a rate given approximately by Stokes's Theorem. Every water globule, being eight hundred times as heavy as air, must be falling through the air, and falling at a rate determined by its weight and size. It reaches a terminal velocity, which it does not exceed, as soon as the propelling force, that is the weight, is equal to the resistance, that is the atmospheric friction. And since one depends on the bulk, and the other on the surface, the terminal velocity for big drops is greater than that for small. Mist globules, or the particles which constitute a cloud, are so extremely small that they sink only slowly through the air. The slightest up-current can sustain them ; so that by the general public they are often thought to float. But every drop of water must sink through the air at its appropriate rate. The large drops of a thunder-shower, and big hailstones, fall with some violence ; though even they may be sustained by strong air currents, and may gradually grow, by condensation, until a hailstone can become as big as a walnut, or, I am told, in some countries as big as an orange— which must be a formidable missile.

A large liquid drop, however, will fall rather slower than the rate given by Stokes's Law for the same quantity of water, because it will tend to flatten out and no longer remain spherical. This flattening out may go on until it breaks up. There must be a maximum size of drop that can reach the ground ; and often a minimum too. As for the very small drops, they seldom reach the ground at all : unless the air is thoroughly damp, they evaporate on the way. For, as Lord Kelvin showed, strong convex curvature promotes evaporation at the expense of condensation ; while concave curvature promotes condensation. Hence it is that in the interstices of woollen and other fabrics, moisture tends to condense, and clothes become damp, before the ordinary dew-point is reached by a flat surface. Hence also the reason why some nucleus is necessary to the formation of a mist globule; for no infinitesimal water drop could resist instant evaporation, and therefore could never form.

It has recently been discovered that a very efficient nucleus for the condensation of moisture is furnished by an electron; otherwise it might seem as if natural precipitation had nothing electric about it. But nature gives many other electric hints. The coalescence of small drops into large ones is assisted by electricity : and the late Lord Rayleigh threw a flood of light on this branch of the subject by numerous and long-continued experiments on the behaviour of drops and jets in the laboratory. Drops must often collide against each other, because, in so far as they are of different size, they fall at different rates. And though they may rebound after collision, yet if there is any perceptible difference of potential between them, they will unite when they collide, on the coherer principle. Thus meteorological phenomena are examples of natural electrical precipitation.

Artificial precipitation is likewise due to the aggregation of particles under electric influence. To bring it about, a brush discharge into air is necessary, and then all particles suspended in the electrified air—whether those particles be solid or liquid—tend to cling together, and are rapidly driven on to earthed or oppositely charged surfaces in their neighbourhood ; so that the air is rapidly cleared of cloud or fume or dust. In cool stationary air, in a laboratory, the process is easy ; but to carry it out effectively and continuously in a violent rush of hot air or vapour, on a large scale, has needed enterprise and invention, and is a notable example of Physics applied to Industry.

A third and shorter portion of this discourse begins to contemplate the possible application of Physics, in the future, to the modification of atmospheric and meteorological phenomena ; especially in certain countries where control of the weather, or regulated precipitation of moisture, seems so desirable as to be worth the trouble and expense of making the attempt.

CONTENTS

PHYSICS IN INDUSTRY

THE first Six Lectures are published in Two Volumes, uniform with the present one, by Mr. HUMPHREY MILFORD, Oxford University Press, Amen House, Warwick Square, London, E.C. 4.

VOLUME ONE.   **2s. 6d.** NET.

Physics and Engineering Science, with Special Reference to Mechanical Engineering. By ARCHIBALD BARR, D.Sc., LL.D., Emeritus Professor of Civil Engineering and Mechanics, Glasgow University.

The Physicist in Engineering Practice. By Sir JAMES ALFRED EWING, K.C.B., LL.D., F.R.S.

The Physicist in Electrical Engineering. By CLIFFORD C. PATERSON, C.B.E., M.Inst.C.E.

VOLUME TWO.   **3s.** NET.

Application of Physics to the Ceramic Industries. By J. W. MELLOR, D.Sc., Principal of the Central School of Science and Technology, Stoke-on-Trent.

The Physicist in the Textile Industries. By A. E. OXLEY, D.Sc., Physicist to the British Cotton Industries Research Association.   *Illustrated.*

The Physicist in Metallurgy. By C. H. DESCH, D.Sc., F.R.S., Professor of Metallurgy in the University of Sheffield.

OXFORD UNIVERSITY PRESS

LECTURE VII

ELECTRICAL PRECIPITATION

DELIVERED IN THE

HALL OF THE INSTITUTION OF ELECTRICAL ENGINEERS
LONDON

On 29th October 1924

BY

Sir OLIVER LODGE, D.Sc., F.R.S.

The Hon. Sir CHARLES PARSONS, K.C.B., M.A., LL.D., D.Sc., F.R.S.,
President of the Institute of Physics,
IN THE CHAIR

PART I

NATURAL PRECIPITATION

PART I

NATURAL PRECIPITATION

EVER since Lord Armstrong's hydro-electric machine, and Faraday's investigation into the details of its working (*Exp. Res.*, vol. ii.), it has been known that drops of water, or mist globules, driven over a solid surface, become electrified ; the sign of the electrification depends on the nature of the liquid and of the solid surface against which it rubs. Dry steam, or dry gases of any kind, will not do what is needful. There must be condensation ; that is to say, the steam must be condensed into a visible cloud, in order that it shall become charged. The issuing steam, when clean, is usually positive, while the jet and boiler become negative.

This, after all, appears only an example, though rather a striking one, of the familiar but notable fact of frictional electricity. Any two surfaces rubbed together become oppositely electrified. And it is well known that the potential which can thus be attained, when the surfaces are separated, can be very high and can give rise to disruptive discharge.

The discovery of electrons has to some extent illuminated this process, but has also perhaps made it more surprising. It is evident that electrons are transferred from one body to another across the junction ; this gives rise to a considerable charge, which rises to a high potential when the capacity is diminished by mechanical separation. When two metals are put into contact, a few electrons flow easily across the junction. Between two insulators an immense number can be forced across, but the exchange is less facile and may require the violence of rubbing to effect it—though whether the act of rubbing is completely understood, may be doubted.

As to the high potential which can be thus attained, each electron is already at so high a potential, at least when isolated, that it is hardly a matter for surprise. Assuming that the ordinary electric laws apply to a charged spherical body so small as an electron, its potential when isolated is comparable to 5000 electrostatic units, or one and a half million volts. And we know how amenable these highly charged and extremely mobile corpuscles are to the slightest electric force or gradient of potential. We also know that the outlying electrons of an atom—those which are responsible for chemical affinity—can readily link together two atoms, so that the link may be said to belong to either, and so that it might be readily transferred from one to the other ; it is, as it were, common property for an instant, and it is almost a toss-up as to which side it ultimately adheres, when the surfaces are pulled apart. Any cause— sometimes quite a slight cause—which encourages a preponderance of transfer in one direction may be said to explain frictional electricity ; a difference in roughness or in colour may suffice, without any necessary difference of texture or substance. If the torn surfaces are really alike in every respect, it seems possible that the intermediate or uniting electron need adhere to neither, but might hover between them in such a state of uncertainty that a puff of air could waft it away. In other words, it might elect to cling to an obtrusive or alien atom, such as air, instead of to the substance to which it originally belonged.

So we are constrained to ask this question : When two precisely similar substances are rubbed together, or are put in close contact and separated, would a blast of air between them become electrified ? The chances are that it would, if the contact were close enough to cause adhesion—as with Whitworth plates. The experiment should be tried.

ELECTRICAL COHESION

In contrast to, or as the converse of, the electrical phenomena produced by the forcible separation or decoherence of two surfaces, we have the correlative phenomenon of the union or cohesion of two surfaces under applied electric stimulus. This is best shown by bringing into so-called, but unreal, " contact " two portions of the same substance ; for instance, two water drops or two mercury globules or two metallic granules. Effective contact is not usually established by any light pressure unless there is some slight electric difference of potential between the surfaces. There is usually a separating film, it may be of oxide, or it may be of grease, or it may be of ordinary air ; and until the separating film is either removed or punctured, cohesion does not occur. For very feeble electromotive forces, below that which is necessary to puncture or squeeze out the film, the surfaces are found to be insulated from each other ; and we have the familiar phenomenon of " a bad joint "—the capriciousness of which has given plenty of trouble to wireless amateurs. It need hardly be said that there is no *real* capriciousness in nature. Caprice is always apparent, not real. (Possibly it is so in human relations too ; but that is less certain.) The law of the bad joint is that for anything below a critical potential difference, it insulates ; while for anything above that, it conducts, until the cohesion is mechanically broken again—a breakage or decoherence which with solids the slightest tremor can accomplish. And inasmuch as any bad joint in practice is liable both to tremor and to electric fluctuation, the reason of its capricious behaviour is obvious, and perfectly rational. A bad joint does not obey Ohm's Law in the least. It is a variable discontinuity, with laws of its own.

The laws of electrical cohesion were investigated by the late Lord Rayleigh, with his usual insight, precise carefulness, and accuracy. He dealt, not with the powder or filings of the ordinary coherer, but mainly with liquid globules, which were more amenable to experiment and calculation. If a small pool of mercury, lying in a flat dish or saucer, is cut in half by a slightly greasy knife, the two halves, though in apparent contact, remain separate. But if the two halves are connected to the opposite poles of a battery, or even a single Grove cell, they reunite into a single pool again. Again, if two jets of water impinge on each other at an oblique angle, they will not unite if the water is clean, but will rebound from each other and continue separate, the two jets being insulated at their place of collision. So also the drops of a fountain of clean water can strike each other and rebound, thereby continuing separate and falling as fine spray. But if a volt or two of difference of potential between the two jets is applied by a battery, they unite and thereafter continue as one. So also if the drops in a fountain or shower are slightly electrified, they too are liable to unite whenever they touch ; and thus they are apt to coalesce into big drops, not falling as spray or fine rain any more, but in blobs like a thunder-shower. This beautiful experiment of Lord Rayleigh's is quite easy to show, with a little care. The jet may advantageously be vertical, about two feet high, and emerge from a smooth glass jet $\frac{1}{25}$ of an inch diameter.

In 1884 I discovered that even the infinitesimal globules of mist or visible steam, if electrified, would unite with each other, so as no longer to be supported by the air, and would fall as Scotch mist or fine rain.

Again, it was found (by Robert Helmholtz, I think), that the light visible cloud of steam issuing from a kettle or other nozzle would, if electricity were discharged into it, appear much darker, becoming brown or orange coloured ; would, in fact, change from the usual light grey appearance, and put on the gloomy and threatening aspect of a thunder-cloud. While, as every one knows, rain, and sometimes very heavy rain, is the natural concomitant of the atmospheric electric disturbance called a thunderstorm.

All this is evidently a case of electrical precipitation, and has remained in many of its aspects a puzzle till quite recent times. There are some things that are puzzling about it even now.

What is the source of electricity in a thunderstorm ? Very high differences of potential are involved, amounting, I suppose, to some millions of volts. At any rate the length of the flashes is enormous. But whence comes the electrical separation responsible for these striking and even alarming effects ?

Prior to 1909 this question could not have been answered, or would have been answered wrongly. It was thought that the charges probably came from the sun ; and most likely charges do come from the sun, giving rise to Aurorae in the Polar regions. Arrhenius has attributed these luminous effects to the magnetic separation of the positive and negative ions, as they fly down from the sun towards the earth. And these electric projectiles appear to be intimately connected with the eruptive solar disturbances known as sun-spots. All that, in one form or another, may be, and probably is, true. But it does not seem to account for the familiar local thunderstorm. A solar effect is on too cosmic a scale. A thunderstorm seems to require a source nearer at hand, and more purely terrestrial, or even local : perhaps something more nearly on the lines of Armstrong's hydro-electric machine, something depending on friction between falling water-drops and the air.

An idea like this may have occurred to many, but it was not easy to see how it could possibly work. If the drops in cloud or rain, or if any mist globules, could be supposed driven against solid surfaces in the upper atmosphere, the difficulty would be removed. But such a supposition seemed absurd. Nevertheless it was found by Lenard that when water-drops splashed, either on to a solid or a liquid surface—when a fountain fell, for instance, into its basin, the scattered drops had a tendency to be electrified. So that in the spray at the bottom of a waterfall, electric separation was manifest. A variety, this, of the hydro-electric effect, and presumably explicable somewhat on the lines of contact or frictional electricity. It was difficult to suppose that the mere breaking up of water into drops would cause electrification. It was thought that the drops must, as it were, rub against something, or else touch and come away, in order to be electrified. It did not seem likely that any friction of water on air would do what was needful for electrification.

Nevertheless the problem of thunderstorms was so insistent that a great meteorologist, Dr. George C. Simpson, took the matter up seriously, investigated the electrical condition of rain, made experiments in the laboratory on falling drops, that is on drops falling through air, obtained numerical or quantitative results, and applied familiar laws of electricity to those results ; so as, in the opinion of many competent judges, to arrive at the theory of thunderstorms. This remarkable Paper was published in the *Phil. Trans. Roy. Soc.* in 1909, having been read in February of that year.

SIMPSON'S THEORY OF THUNDERSTORMS

The remarkable thing discovered by Dr. Simpson is the hitherto quite unexpected fact that the mere breaking-up of a drop by a current of air results in electric charge. When a large drop, falling through air, breaks up into small ones, the air goes away negatively charged, the water positively charged, the water having touched nothing but air. Experiments made previous to 1908 to detect such an effect had given negative results. Probably the experiments had not been very persistent, since such a result could hardly be expected on theoretical grounds. It came as a surprise, and has taken some time to be assimilated. Some chemists, among them Professor H. E. Armstrong, have indicated disbelief, or, at any rate, serious doubt. But, so far as I know, no experiments have been made to contradict the result ; and I think it must be taken as substantiated. At any rate I propose so to take it. But surely it is extraordinary that the mere breaking-up of water should electrify it.

To attempt to understand it, we must look at it from the point of view of what I began with—namely, the electrical influence on cohesion. It may be regarded as the reciprocal or converse effect. Two globules unite under electrification ; and two globules separating give rise to electrification. But whereas the combination must be a differential effect—that is to say, there must be some difference of potential between the two globules in order to break down the barrier, the separated portions of a single globule are not differentially electrified. Both portions are positive, and may apparently be at the same potential. The opposite charge in their case belongs to the air, which has, so to speak, blown them apart. A large drop of water, falling through air, first tends to flatten itself by the viscosity resistance, and then presumably crimps itself into drops all round its horizontal circumference. Into how many drops it may break up I do not know, nor does it seem to matter. What is certain is that a big drop breaks up into smaller ones, and these, according to Simpson's experiments, are positively electrified with reference to the air.

The small drops naturally fall at a slower rate than the big one, and accordingly if there is an uprush of air, that is if they are falling through an ascending current, they will be carried upwards. But being electrified, they are just in a condition to cohere again as soon as they come into contact. For they can hardly all be exactly alike ; they may differ either in size or in potential or in both, and Lord Rayleigh found that any difference, whether in size or in potential, facilitated recombination. Hence, sooner or later, the small ascending globules will coalesce into big ones, and then they will be able to fall again, once more to be broken by the up-current, and so carried up again, this time with an additional charge. This is Dr. G. C. Simpson's theory in brief ; and he has adduced many meteorological facts which appear to justify it, and to show that the imagined process is one that is really likely to occur. The occasional sustaining of heavy matter by air currents is well known, and the production of large hailstones can hardly be otherwise explained. The sequence of operations : (1) coalescence of mist globules into falling drops ; (2) breakage of large drops and carrying upwards ; (3) renewed coalescence and falling again ; may be repeated several times, until presumably the drops get so highly charged that they will no longer combine and fall. For, be it noted, though gentle electrification combines them, strong similar electrification will keep them apart ; for they will repel each other, and so never come into contact. Recombination can only occur when they do come into contact. Hence when they are too highly electrified, they will be carried by atmospheric uprush to a higher stratum, where they can form part of a cloud. When they arrive among other uncharged mist globules, they can easily combine with them, and so once more grow into bigger ones, and fall. But if the uprush is sufficient, they will not reach the ground ; they will be broken up again and recharged. And so the process will go on, until, by some horizontal drift, they escape from the up-current, and find a region where they can fall, even as big drops ; thus constituting a thunder-shower round the edge or at the margin of the uprushing air.

There is thus a certain critical speed for an air uprush which will prevent any water from being able to fall through it. For small drops will clearly be carried up, and big drops will be broken into small ones. Hence no rain can fall through a sufficient uprush of air. The speed of air which would prevent water from falling through it is nothing excessive. The critical speed had already been measured by Lenard, and was found to be 8 metres a second, say 26 feet a second, or 17 miles an hour. This, as Simpson says, is but a moderate wind ; and, in storms, uprushes are known to be able to lift solid objects—showing that they are much more violent than the critical speed.

Discussion of the Theory in the Light of Cohesion

Now let us see if we can form any conception of what is happening when a drop breaks up under the influence of a vertical wind. To take a simple case, let the drop break into two. The two halves will be connected by a ligament, under the influence of surface tension ; and when the length of that ligament is equal to its circumference, it becomes unstable, and snaps in two or more places, while its middle coalesces into a very minute drop or drops.

Now cohesion is known to be a form of electrical attraction. We may conceive of the two halves being held together by a surface layer of electrons, or what is called a double layer, straining across the gap. The electrons are more mobile than their positive correlatives : and when the snap occurs, it seems possible, though not easy, to imagine that electrons may be carried away by the air current. Whether the surface electrons are themselves blown away, or whether they crowd into the little drop which is formed by the broken filament, I know of no evidence to show. Anyhow they do appear to be carried away. Otherwise the remaining drop could not be positively charged. The air carrying off the electrons is thereby negatively charged ; for even if the electrons are not isolated and bare, but belong to a small drop, that drop, being small, would almost instantly evaporate, with the ultimate result that the electrons are practically free ; or possibly they attach themselves to the molecules of dry air, giving the known phenomenon of a charged gas.

Thus is effected electrical separation. The negative goes up, the positive stays down or partly down, until the vertical potential gradient rises to a critical value, say 30,000 volts per centimetre ; which is well known to be sufficient to give a disruptive discharge or flash between two flat conductors. Probably, under the circumstances, a much less potential gradient than that would suffice — but something perhaps not hopelessly below that general order of magnitude.

Quantitative Estimate and Extension

Now let us look at the matter more quantitatively, making use of Dr. Simpson's excellent measurements. The drops he mostly experimented upon were a quarter of a cubic centimetre in volume, and therefore, when spherical, would have a radius of four millimetres. Such a drop, falling on a vertical jet of air, is blown to fragments ; and the fragments, being collected, are found to have an aggregate charge of 0·005 electrostatic units. To provide such a charge it is easy to calculate that ten million electrons must have been removed from the quarter cubic centimetre of water, or say forty million per cubic centimetre. This seems a large number ; and indeed it continues to be surprising that an air current can effect this separation. Still, that is what the facts assert : and, after all, considering the number of atoms in a drop, ten million is a trifling number.

Let us think where they would come from. They would be on the surface, they would be, in fact, on the broken surface, let us say on the section of the filament when that snaps. So we can reckon the size of that filament which would suffice to supply ten million electrons in its cross-section, if each atom in the cross-section contributed one electron. The atoms in the stratum are 10^{16} per square centimetre. So the area of cross-section, in order to supply 10^7 electrons, would only need to be 10^{-9}. That is, its diameter would be 3×10^{-5}, which is the thirty-thousandth of a millimetre. This is improbably small ; but then it may be unreasonable to suppose that the cohesion of the water is effected, at breaking-point, by one electron from every atom, and it is still more unlikely that every such electron would be blown away. If only one in a thousand of the cohesion-electrons were blown away, the cross-section needed would be 10^{-6}, and the diameter the hundredth of a millimetre : which is still very small. Hence, as far as quantities are concerned, there seems no difficulty about supplying the measured charge in this way.

Discussion of the Potential likely to Result

To go further into the matter, it will be well to attend to differences of potential, and to the tension across the separating film. Go back then to the split mercury globule or pool, with the two halves resting against each other, but not in contact, and consider the film separating them as something rather thicker than molecular dimensions, say, for instance, 10^{-7} or 10^{-6} of a centimetre. Lord Rayleigh showed that often one Grove cell could break down the film and effect junction ; let us say 1·5 volt. Or perhaps for numerical purposes it would be better to take it as one volt ; since with care it can be brought down to that, and at any rate it is of that order of magnitude. Taking the film as 10^{-6} cm. thick, the gradient of potential is a million volts per centimetre. This will equal $4\pi\sigma$, where σ is the surface density of charge. Hence σ is of the order 300 E.S. units, and the number of electrons which would give this surface density is 6×10^{11} per square centimetre. Or, in the section of a filament one-hundredth of a millimetre thick, 6×10^5, which is getting near the right order of magnitude for providing the 10^7 electrons required, as estimated two paragraphs above.

A film of thickness 10^{-7} cm. would give tenfold the density for the same minimum voltage, and ten times the number of electrons ; but the film in that case would be of black-spot thinness.

The tension or mechanical force across the film—even a film that gives colours of thin plates—is considerable, being $2\pi\sigma^2 = 5 \times 10^5$ dynes per square centimetre ; say half a kilogramme per square centimetre, or 7 lb. to the square-inch, even for 1 volt. It appears probable that it is this mechanical pressure which forces the drops into contact, squeezing out the residual film between the surfaces, or, at any rate, squeezing it out in some one locality. For anything which promotes irregularity in the film, such as fine dust, promotes cohesion, by concentrating the tension in one or a few places, giving easy opportunity for the film to accumulate in local pockets without having to go far. And, of course, directly cohesion sets in at one point it rapidly spreads.

I regard then the breaking-up of the drop, and its simultaneous electrification, as the reversal of this cohesion process ; and I look for a charge corresponding to the surface charge of the broken filament, which had been held together by a surface density of the order above reckoned.

It might seem likely that the steep gradient of potential across the film would give rise to puncture by disruptive discharge. But Lord Rayleigh adduces arguments against that view — though it certainly was a reasonable view, and may be legitimate as an alternative to that of squeezing together by hydrostatic pressure.

For our purposes it does not much matter whether the union is due to static pressure or disruptive discharge. All we need is the electrons straining across the gap, or across a surface which was a gap before reunion, or which becomes a gap directly the union is broken.

Once the separation is effected, the potential of course rises. The observed charge, 0·005 units, given to the original falling drop of four millimetres in radius, would raise its potential about four volts. The same charge shared among a multitude of drops, say n^3 in number, would only raise the potential of each to $1/n^2$ of that value, their linear dimensions being $1/n$th of the original. For instance, if the big drop broke into eight small drops, the radius of each would be two millimetres and the potential of each would be one volt. That is to say, we are still within the right order of magnitude for the potential which effects coherence, and therefore presumably within the right order of magnitude for the potential resulting from disruption.

Hence these theoretical and quantitative considerations tend to justify the rather surprising hypothesis on which Dr. Simpson originally based his theory and his experiments ; namely, that the breaking of drops of water by an air current could give rise to electrification, and by repetition under reasonably likely atmospheric conditions could even generate the violent electric discharges of a thunderstorm.

PART II

ARTIFICIAL PRECIPITATION

The following references may be useful :

Nature, 26 July, 1883, vol. xxviii. p. 297.   Preliminary Letter.

Phil. Mag., March, 1884.   Paper on " Dust."

Nature, 24 April, 1884.   Lecture to Royal Dublin Society.

Nature, 22 January, 1885.   Lecture to British Association, Montreal.

Engineering, 5 June, 1885.   Paper by Mr. Alfred Walker.

Proc. Roy. Inst. of Great Britain, 28 May 1886.   " On the Electrical Deposition of Dust and Smoke, with special reference to the Collection of Metallic Fume, and to a Possible Purification of the Atmosphere "; also

British Association Report for 1885, pp. 744 *et seq.*   " Electrostatic View of Chemical Action."

PART II

ARTIFICIAL PRECIPITATION

ARTIFICIAL precipitation, which is an interesting application of static electricity, dates for practical purposes from an observation which the late Mr. J. W. Clark and I made at Liverpool in 1884, though it was found that something of the same sort had been casually observed by a Mr. Guitard, and mentioned briefly in *The English Mechanic* of 1850. Indeed, I have just learnt, from an excellent article, " Precipitation Electric," by Mr. H. J. Bush in Thorpe's *Dictionary of Applied Chemistry*, that a certain Hohlfeld of Leipzig made a similar observation so long ago as 1824 ; so that this year is its centenary.

General Dust Phenomena

Mr. Clark and I began by an inquiry into the phenomenon of the dark plane or dust-free space rising from hot bodies, which is made visible by letting the air stream upwards into a horizontal beam of light. Dust in the air is said to make the beam visible ; a fact which is more truly expressed by saying that the beam makes the dust visible. So if from any portion the dust is cleared away, leaving a dust-free space, there will be nothing to see in this portion. It will accordingly be quite dark, in contrast to the rest of the luminous matter surrounding it, and accordingly has the appearance of dark smoke. A similarly misleading appearance can be noticed when smoke is escaping from a room full of smoke through a window in daylight : the smoke looks as if it were coming in, because the fresh air blowing in looks much darker than the rest. Similarly, a spirit-lamp flame held below an electric beam looks as if it were smoky ; but a hot poker does just the same. And Tyndall showed that the dust-free space had nothing to do with combustion or the burning up of the dust—though he did not arrive at the true solution.

Lord Rayleigh began to examine the phenomenon more critically, and found that quite a small excess of temperature, such as that of hot water communicated to a rod, would suffice to give a dust-free plane of singular definiteness and regularity, rising as a lamina from the rod.

Mr. Clark and I continued the investigation, and found that a warm rod was surrounded by a dust-free coat, which coat continually rose from it, so as to constitute the plane, and was continually renewed. All which was described in the *Philosophical Magazine* for March 1884, with illustrations of the effect. It appears that small dust particles cannot get into contact with a hot body, or even a warm body, but are bombarded from it by molecular impacts—very much on the lines of Crookes' Radiometer; the force being more effective on the minute particles of dust than it is on the large vanes of the Radiometer.

The late Mr. John Aitken also took the subject up, and in many interesting ways showed that a warm surface would keep itself clear of dust, while a surface colder than the air would have dust deposited upon it. Molecular bombardment, somewhat on the lines of Brownian movement, drives the dust from a warm surface and towards a cool surface. So that whenever warm air streams near a wall, the wall gets blackened, as if the air had been smoky—giving to the wall or other surface a dirty appearance quite familiar above stoves and gas flames, and even above electric-light bulbs, out of which naturally no smoke or anything else can emerge.

After working at the subject in the autumn of 1883, Mr. Clark and I went on to examine the matter more closely with microscopes and other appliances; and, among other experiments, we electrified the rod to see what effect that would have. We found to our surprise that the dust-free coat thickened, and expanded, and rapidly extended into the box; in other words, that the whole box was cleared of dust by the electrification.

I then proceeded to electrify smoke, not on a small careful metrical scale by means of a battery, but in a larger and more violent way with a Voss or Wimshurst machine. The appearance was very striking. The smoke particles aggregated together like filings round a magnet, hovered in the air, after the fashion of the piece of gold leaf called "Mahomet's Coffin," for a short time, and then clung to the floor and the walls of the vessel, the effect being particularly rapid and efficient when brush discharge took place from a point. I also filled a bell-jar with steam, that is, with a visible cloud, and found that electrification caused the ultra-microscopic drops to cohere together, and fall as fine rain or Scotch mist.

I gave one of the Evening Lectures at the British Association Meeting at Montreal in August 1884, on "Dust," and there showed this effect on a fairly large scale, to the delight of both Lord Kelvin and Lord Rayleigh, who were present on the platform. This, one may say, was the beginning of artificial electric precipitation.

PRACTICAL APPLICATIONS

Soon afterwards, Mr. Alfred Walker, of the firm of Walkers, Parker & Company of Chester, tried to apply the effect on a large scale in his smelting works at Bagillt in North Wales, where a quantity of lead dust escaped into the atmosphere, to the damage of the neighbouring agriculture. But the method of producing high-tension electricity in those days was rather primitive, and by no means of an engineering character. The difficulties of insulation were not properly appreciated by him. It is doubtful if any real electrification was communicated to the flues, along which the hot flaming and smoky gases were rushing at a considerable pace from the smelting furnaces: so that the first attempt at practical application was unsuccessful, and I presume was discontinued.

Some years later, however, the attempt was made again, after many large-scale laboratory experiments at Liverpool, by my son, Mr. Lionel Lodge. By that time the vacuum valves, which had been improved by myself and Mr. Robinson, enabled the discharge from an induction coil to be rectified ; so that continuous high-tension electrification could be maintained from an alternating or intermittent dynamo current and transformer, in a more satisfactory and engineering manner than by an electrostatic machine, such as had been used by Mr. Walker. Moreover, the kind of electrodes most suited to the apparatus became known, after many experiments, and special elaborate insulators were constructed to hold them. It was found that the best results could be obtained by suspending oppositely charged metal surfaces alternately in a large precipitation chamber, alternate ones being provided with point or edge dischargers. There were difficulties connected with the clogging of the points or edges by the dust, and many subsidiary contrivances were devised before the arrangement became finally practical.

Meanwhile Dr. Cottrell, in America, had been working on somewhat similar lines, and had begun to apply the process to smelting and other work on a really large and successful scale. The Lodge Fume Deposit Company was also started in this country, and the two firms decided to amalgamate and co-operate, Dr. Cottrell always acting in the most honourable and friendly manner. His enthusiasm carried people along, and greatly helped the spread of the knowledge of the device in America.

I am told that a Dr. Moeller, in Germany, has independently taken out a number of patents, and that some of them are being applied on a large scale by the Metalbank Metallurgische Gesellschaft, who are also in friendly co-operation and have agreed to delimitation of frontiers.

During and since the war the process has been applied on a very large scale to take the dust out of blast-furnace gases, at four ironworks ; the work proceeded under excellent facilities, because the dust was full of potash, which the Government required. The dust removed from blast-furnace gases is therefore of some value, a value which may be wasted, but which ought to be developed ; but the chief advantage of removing it lies in the fact that dust-free gas burns much better, and therefore is more efficient for heating the hot-blast or for generating steam, without clogging of the flues or interfering with effective combustion and doing other damage. I may remind you that dust is actually used in collieries to prevent combustion and the spread of an explosion. So when you try to burn dusty blast-furnace gas, the flame is continually hesitating and retreating or going out, and often will not light ; but when freed from dust the flame roars splendidly.

The recovered dust from some ores contains 20 per cent. of potash, and should be of considerable agricultural value ; but it may also contain cyanide, which is deleterious ; this, however, ought to be capable of removal. The quantities of dust to be dealt with may amount to a hundred tons a week. Blast furnace dust is singularly light, and therefore troublesome, as deposited : a ton of it more than fills a 10-ton truck. In an installation in which 20 tons a day are deposited, the gases rush past the electrode in volume 50,000 cubic ft. a minute.

I suppose blast furnaces represent the largest kind of installation in this country, since the quantities dealt with are so great. I must say I admire the engineering skill which enables continuous high-tension electrification, at a potential of nearly a hundred thousand volts, to be applied continuously night and day in enormous chambers to rapidly moving hot gas, with very little attention ; so that some 95 per cent. of the dust is deposited, and so that tons of the solid material shaken off mechanically from the electrodes and collected in channels or chutes, are deposited every day and thrown down into railway-trucks below, and carried away. I am told that, in all, there were two hundred installations at work in the world a year ago.

Another installation, I may mention, is at the tin-smelting works near Bootle, north of Liverpool, belonging to Messrs. Williams Harvey & Co., where the tin-oxide fumes which formerly escaped into the atmosphere and constituted a nuisance are now saved, along with the smoke and other solid material from the furnaces, thus effecting great economy and paying for the electrification again and again. It is easy enough to re-smelt the recovered material ; and thus a lot of valuable metal is saved, which would otherwise have been wasted.

PRACTICAL DETAILS

Detailed information about some few of these installations is contained, with illustrations, in an Appendix. I need only say here that I am fairly well acquainted with some of the difficulties encountered by my sons in making large-scale application, and with some of the devices they have adopted in overcoming them. I will only refer to a few of these.

First of all the points or edges, and indeed all the surfaces, tend to get clogged with dust ; so that periodically (about every three or four hours) everything has to be hammered or vibrated in some way. It is found best to knock the frames from below, by considerable weights hanging from levers above, which can be worked by hand, like railway signals ; or else to arrange the knocking to be done automatically. The lighter hammering of the insulated and charged portions must, of course, be effected without earthing them ; and that is done by a kind of projectile thrown or knocked up by the larger weight, which itself only hits the massive portion of the plant—the portion at zero potential.

Another feature where gases are treated at comparatively high velocity is that the dust collected on the surfaces tends to get carried along mechanically by the gases, unless pockets are provided in which it can accumulate—which it does mainly by means of eddies.

Another and more serious difficulty is the electrical surgings, which are always liable to happen from a large charged capacity, which surgings may damage insulation and give other troubles, but these are now understood and provided against.

Again, it is found that vacuum valves, though efficient enough on a small scale, are not suited to heavy engineering work ; revolving commutators, synchronously driven by the alternating current supply, are used either to supplement or to replace the glass valves.

When everything is established and properly arranged, it is remarkable how little attention is required. The whole thing is set up in units, and each unit requires its separate periodical hammering ; but short of that, and the removal of the dust, there is little or nothing to be done. The electrical appliances are so designed that they carry on without interruption throughout the year.

Naturally great precautions have to be taken to secure the safety of the workmen ; no unit can be entered without unlocking the door, the key of which hangs on a hook, so arranged that when the key is taken off the hook, that unit is earthed and so put out of danger, without stopping or incommoding the electrical supply.

The installations are mostly of two kinds, " plate " and " tube." In a plate installation, the earthed portion consists of large flat or corrugated plates, with wire or comb electrodes between them—all, of course, elaborately insulated. In tube installations, the earthed portion consists of vertical tubes, the diameter being chosen to suit particular gas conditions. Down the axis of each a thin uncovered wire, say number 16 gauge, is hung from an insulator and kept stretched by a weight below.

To illustrate some of the curious unexpected difficulties which can be encountered and easily overcome when detected (though the detection is not an easy matter), it may be mentioned that in a sulphuric-acid chamber the insulation was constantly proving defective, in a way which could not be traced to any defect in the insulators. It was ultimately found that this depended on the shape of the weights which stretched the insulated and discharging wires. That shape had not been particularly attended to ; for it would seem that any weight would serve to stretch the wire. But, in practice, a thin stream of acid flowed from the bottom of these weights, and thereby earthed them. The difficulty was readily overcome, when detected, by giving them flat bottoms or grooving them, so that they should drip instead of running as a continuous stream.

In some of the installations the gases are worked and the dust precipitated at a red heat : for instance, in the precipitation of iron oxide and dust from SO_2 gases generated by roasting pyrites for the manufacture of sulphuric acid by the Chamber process. The heat of the gas is utilised in Glover towers so that the dust has to be removed from the hot gases as it leaves the burners, and it is found that there is no necessity to cool it down. Subsequently further treatment of the cooled gases can be arranged to deposit sulphuric acid mist and impurities.

The sign of charge used in all cases is usually negative, the positive being sent to earth. The rate at which the gas streams past the electrodes is of the order of 7 ft. per second, but, of course, it varies in different cases.

By this time a good deal of experience has naturally been obtained. Different devices have to be used when dealing with different materials, and without experience it would be uncertain what kind of arrangement was most suitable for any particular case. Some materials are undoubtedly easier than others to deal with. Iron oxide seems particularly easy. Tin and zinc and other metals are tractable enough. Lead, for some reason, seems more troublesome than others—the oxide is apparently too good an insulator and declines to receive a charge readily. It seems remarkable that sulphuric acid can be dealt with without undue difficulty, but I found that this could be done, at any rate on a laboratory scale, so long ago as 1888. (See a letter of mine in *The Electrician*, dated 5th January 1889, from which the following extract is made :

"... So far as preliminary small-scale experiments on different kinds of fogs are concerned, they have all been done. Beside chemical smokes of various kinds, I have dispersed steam, and turpentine smoke, and coal smoke, and sulphur, and mixtures of all of them. I have made stuff like that in a sulphuric-acid chamber by burning sulphur and supplying nitric acid vapour, and steam from a boiler. All manner of nameless abominations have been tried, and all are amenable to the electrical influence.")

Smokes of Towns and Improved Combustion

It has often been suggested that this method of electrical precipitation of smoke and dust might be applied to the atmosphere of large towns. But the difficulties of applying it in the open air are very great ; and it has never seemed to me the right method of dealing with town smoke. Smoke is an extravagant thing to produce ; and it would be expensive, as well as dirty, to deposit it on the houses and people of a town. The right method of dealing with town smoke is not to produce it. Its production should be avoided by improved methods of combustion. Unfortunately, with ordinary methods, when water has to be heated, or steel slabs like armour-plate have to be annealed, smoky flames and smoke-laden spent gases are more efficient than those which are dust-free. For (1), combustion cannot go on in contact with relatively cool surfaces, like those of a boiler or armour-plate, and so inevitably the heat has to reach such surfaces across a gap by radiation ; and (2), clean air is a bad radiator, while smoky air is a much better one. But still it cannot be considered a satisfactory method to employ smoke for that purpose, any more than it is satisfactory to depend on the carbon in a gas-burner flame for the radiation of light. The scientific way is to realise that radiation is necessary, and to provide red-hot solid radiators—like highly magnified gas-mantles, which, in a clean and non-smoky atmosphere, can propel the heat through the ether whither it is desired. Contrivances for this radiant heat method of heating boilers have been devised and applied, by Professor Bone, in a manner which, if not as yet entirely successful (and on that I express no opinion), seems to me to contain the germ of the proper method of heating moderately cool surfaces, namely, by specially arranged and controlled solid radiation. The transmission of heat by mere convection from hot air, after flame has subsided, is not very efficient, especially if surfaces are clean ; but a black or dirty surface absorbs radiation perfectly well.

The other alternative (which doubtless has difficulties of its own) is to use conduction ; that is to say, to protrude from the boiler, as part of its construction, rods or flanges long enough to extend into the flame and there to become red or white hot at one end. In that case combustion is not interfered with, the flame can touch the hot metal, and the heat thus received by solid metal would be freely conducted into the boiler. The gradient of temperature necessary to drive heat along a rod of metal is very much less than the discontinuous step of temperature needed to propel it across a film of air or of oxide, or any other insulating film or stratum. The difference of temperature, between the flame of a furnace and the boiler-plate which has to absorb the heat, is very great, maybe as much as a thousand degrees centigrade, or nearly two thousand Fahrenheit. It must be admitted that this drop of temperature is inefficient—as seriously inefficient as an unused waterfall : high-level heat is falling to low-level without doing any work, without achieving anything but its own transit. The *heat* need not be wasted ; it may be got to enter the boiler, but great water-tube surface is required, and, if conduction could be utilised, the flow of heat from a flame into the boiler could be made much easier.

This is a digression from the subject of electrical precipitation. But the difficulty of boiling water without smoke is a real one ; and the inefficiency of the temperature drop is deplorable ; so in speaking to the Institute of Physics, I allow myself to refer to the subject once more.

It used to be thought that the incoming of the gas engine, where furnace temperatures exist within the actual working cylinder, would have the effect of dispensing with a great deal of steam production, and would thus get rid of this source of inefficiency—though the fact that the cylinder has to be water-cooled evokes another kind of inefficiency, for there again heat is flowing downhill without doing any work. But though large gas engines have arrived, and though a hopeful combination of gas and steam engine has been devised, in the hope of turning some of those difficulties into advantages, the invention of Parsons' steam turbine, one of the remarkable achievements of our time, has given to the utilisation of steam alone a new lease of life. For, although the production of steam is still an inefficient process, the utilisation of the energy in the steam by the modern turbine is about as efficient as can be imagined. Moreover, the device is so convenient and manageable that it is almost inevitably adopted : so that attention to the steam-raising problem becomes once more an alive and important necessity.

All these things may be superseded when the time comes for the utilisation of atomic energy, but that time is not yet. Whether our grandchildren will live to see it, on anything like an engineering scale, may be doubted. Time will show.

PART III

COMBINATION OF THE TWO; WITH SUGGESTED
METEOROLOGICAL POSSIBILITIES

PART III

ARTIFICIAL METEOROLOGICAL POSSIBILITIES; OR COMBINATION OF NATURAL AND ARTIFICIAL PRECIPITATION

I HAVE now spoken of the natural and of the artificial kind of precipitation. But the modern fashion is not to leave Nature alone—rather to encourage and supplement it. Hitherto we have left large-scale atmospheric processes severely alone, and—as used to be done with diseases, plagues, and other ill-understood visitations, we have not attempted to coerce or control Nature, but have taken refuge in petition and appeasement of Higher Powers, in the hope that they will do what we have not the spirit or the energy to accomplish. Fortunately Pasteur has lived, and diseases have been taken in hand with some measure of intelligence and knowledge, and with results which are already profoundly moving. Doubtless much more remains to be done in this direction.

The Pasteur and Lister of the atmosphere have not yet arrived. Accordingly, though we have learnt that the precipitation of moisture depends on electrical conditions, and though rain comes down electrically charged, in a way which ought to give us a hint, we still supplicate Higher Powers for the production or the limitation of rain, instead of setting to work to see what we can do for ourselves.

I admit that the problem is a large and difficult one. But so are all problems, until we begin to tackle them. The atmospheric difficulties, however, are peculiar in this : they are on so large a scale that no ordinary laboratory experiments, or anything within ordinary private means, suffice even to make an experiment. Experiment is therefore left, in so far as it is conducted at all, to people—the so-called rain-makers—with some wit for influencing and interesting their fellow-citizens, and with some hope (so it appears to me) that the element of luck will intervene in their favour. The only thing that can be said is that any experiment is better than none, and that occasionally even what Darwin used to call a " fool-experiment " suggests a clue or bears some sort of fruit. But experiments conducted with more meteorological knowledge would surely be better : and now that it is possible to produce high-tension electricity on an extensive engineering scale, it seems to me that something ought to be attempted. Whether the stopping of rain or the production of rain is the easier problem, I am not sure ; but what the greater part of the world suffers from is drought.

Now, if there are no clouds, or extremely little moisture in the atmo-

3

sphere, the case is hopeless. We cannot assume that any cause will put moisture into the atmosphere, except the sun. But I am told that in the countries suffering from drought, clouds do at times accumulate, but disappear without precipitation. Why should they not be electrified ? Or, since they are probably already electrified, why should not the sign of the electrification be changed ? Or again, why should not one part of a cloud be electrified differently from another part, so that the drops in it should be of different potential, and be likely to run together and coalesce ? In other words, why should not natural precipitation be assisted artificially ?

I know there are many difficulties. They may turn out to be insuperable ; but that has not usually been the way with difficulties before, and we cannot tell what is insuperable until we try. The experiment would be costly, it would probably fail at first ; there is much to learn, but there is also a good deal of knowledge to guide us ; it would not be entirely working in the dark. Sooner or later the experimenters would gain a clue, some hint which might put them on the right track. They would very likely not start on the right track : they would have to mend their ways. They might encounter some ridicule ; they might be discouraged by failure. But I feel that success awaits those, possibly of a future generation, who can put up with ridicule and who are not depressed by failure, but who persevere and overcome obstacles—which, after all, when we come near them, are seldom found as insuperable as we thought.

There were difficulties in navigating the Arctic Ocean, but Nansen overcame them in a pioneering manner. There were difficulties about exploring the Antarctic continent, but Scott and Shackleton and other noble men gave their lives in the attempt. There were difficulties in scaling the Himalayas, they seemed an insuperable obstacle. Whether the summit of Mount Everest has been reached by those two who died on one of its ledges, I do not know, nor do I know what result is to be expected if it has. But men will do these things ; they do not seem to count the cost. They press on with enthusiasm, and leave the results to posterity. There were difficulties about making the equatorial belt of the earth habitable for white men, but Ronald Ross and the others worked at it—with what results we know.

There are difficulties about the electrical controlling of the atmosphere. Is that to be the one region of the earth over which man has no power, and about which he must succumb supinely to fate ? I do not believe that it is. I feel sure that if the control of the atmosphere is felt to be an important problem, it will be tackled either now or by posterity. The considerations to which I have called attention in this lecture are such as to give some kind of clue, perhaps only the inkling of an idea. There are meteorologists who know far more about the atmosphere than I do. They will, I expect, be conservative in their estimate. It may be that physicists rush in where meteorologists fear to tread ! But anyhow the problem strikes me as no more difficult than the problem of disease at one time appeared ; and I venture to regard the future with hope.

APPENDIX

PRACTICAL NOTES ON COMMERCIAL PRECIPITATION

(Drafted by Lionel Lodge)

APPENDIX

IT is difficult to give general information, as each specific purpose usually requires particular treatment. As far as possible, certain types of plants are standardised for particular purposes, only varying in capacity and control arrangements ; but variations in the temperature, speed, or content of the gases may involve considerable alterations in the precipitator chambers, and this again may affect the potential necessary, insulator design, methods of dust removal, and so forth.

The accumulated experience from over two hundred plants makes it possible (given information as to composition of gases, temperature, rate of flow, nature and quantity of suspended particles) to determine the most suitable type of treater for practically any fume problem, and to calculate closely the efficiency obtainable, power consumption, drop in temperature, etc.

The different uses to which the process has been applied may be grouped under the following headings : Acid fumes ; waste gases from metallurgical processes ; combustible gases ; air cleaning ; miscellaneous dusts.

Publications have been made concerning particular installations for all these different uses, so only one or two examples will here be given.

SULPHURIC ACID PLANT

In the manufacture of sulphuric acid from pyrites or spent oxide, for instance, the process is used to remove the dust from the hot sulphur dioxide gases as they leave the burners. The dust, which consists largely of oxides of iron, is of no value, but constitutes a nuisance in the subsequent processes, as it tends to choke up the Glover towers and discolour the acid.

When sulphuric acid is made by the contact process, it is essential to remove all traces of dust and other impurities likely to injure the catalyst. Two separate treaters are employed. One deals with the hot gases as they leave the burners at a temperature between 450° C. and 600° C., where the dusts and any solid particles are removed. The other deals with the gases when cooled. The precipitator at this point removes traces of arsenic and other impurities as well as precipitating any acid mist, the operating efficiency of this process in actual commercial use being over 99 per cent.

A type of installation for removal of the dust is shown in Fig. 5. This plant consists of horizontal twin chambers carried on a steel framework, the walls being built of brick 18 in. thick to avoid, as far as possible, loss of heat. The gases enter from the burner house on the extreme right. The overhead main can just be seen entering the valve tower near the top. The gases pass downwards through cast-iron valves 3 ft. in diameter, then between the collecting electrodes, which, in this type of chamber, are made of heavy-gauge wire mesh hung vertically and suitably stiffened to maintain a flat surface ; this construction is used on account of the high temperature, the whole of the inside being more or less red hot. These collecting electrodes are arranged in four separate banks, each bank consisting of six electrodes equally spaced across the chamber so that there is a free passage between them for the gases, which pass over, not through, the surface of the electrodes. Midway between each pair of collectors are hung the discharge electrodes, consisting of a plain fine wire attached to suitable fittings to facilitate quick erection and easy renewal. These electrodes are spaced about 6 in. apart and supported on frames, top and bottom, to hold the electrodes taut, the frames being carried on high-tension insulators. Inspection doors are arranged immediately below each bank of electrodes. These are conspicuous on the photograph. Below these doors are the dust hoppers with mouth-pieces and slides to empty the dust into the trolleys running on rails below.

Each half of the twin chamber is identical with the other. Normally the gases pass through both sides, but either side can be shut off for removal of dust. During this period all the gases pass through the other side. Removal of the dust is a simple and quick operation, taking only a few minutes. The valves at each end of the chamber are closed and the hoppers opened ; most of the dust is immediately emptied out, but rapping gear is provided to tap the electrodes slightly in order to remove any dust which may be left adhering. The hoppers are large enough to carry two or three days' collection of dust, so that the time spent in looking after this plant is negligible. In the ordinary way it works with little or no attention.

The speed of the gases varies, but 7 ft. per second may be taken as an average velocity. Uni-directional current is supplied at a potential of about 60,000 volts, the power consumption being about 3 kilowatts for plant to clean the gases from a pyrites burner of 18 to 20 tons per 24 hours capacity.

Tin Smelting Plant

A plant erected in this country about three years ago and treating the waste gases from a Tin Smelting Works may be taken as a typical example of a precipitator for dealing with fumes from metallurgical processes. It employs pipe precipitators, the pipes being 9 in. in diameter, with 48 pipes to each unit ; there are fourteen such units. The lower part of each unit consists of a ferro-concrete box in which the dust is collected. This box is some 12 ft. square and about the same height. The gases enter near the top, then pass downwards round the pipes and upwards through the pipes into the top header, and away. Large cleaning-out doors are provided for periodically raking out the collected fume of which there is considerable bulk. This fume used to escape from the stack in the form of black smoke, and constituted a serious nuisance in the district. Since the precipitator was put in operation the visible discharge from the stack is merely a wisp of steam. The weight of dust collected amounts to 3 or 4 tons a day. This dust carries a fairly high tin content. The process is continuous day and night, being shut down only once a year for general cleaning.

The Transformer House equipment of five units is arranged so that any precipitator can be connected to any transformer unit, each of these units having nominal 25 K.V.A. capacity. The transformer and all high-tension parts are elaborately screened, so that it is practically impossible to approach any high-tension conductor while the current is on. Automatic safety switches earth each section when any door is opened, and the circuit breaker on the primary side of the transformer cuts off the current. In the pre-cipitator house all the doors are locked and controlled by a master key, so that before any door can be opened the key has to be obtained, and this automatically earths that entire section. The operating pressure of this installation is approximately 80,000 volts.

Blast Furnaces

The cleaning of blast furnace gas is an example of precipitators in the third group (combustible gases). These installations handle a large volume of gas amounting to several million cubic feet per hour, and the dust collected, which is extremely fine and light, amounts to many tons per week.

The installation shewn in Fig. 6 has a capacity of approximately 4 million cubic feet per hour at working temperature, and handles the gases from two blast furnaces. There are eight separate precipitating chambers. The gases enter on the far side of the building from a 6 ft.-diameter main, then pass straight through the chambers, which are about 30 ft. long, and enter the delivery main. The collecting electrodes in each chamber consist of narrow plates, about 10 ft. long, spaced about 9 in. apart. Between each pair are the discharging electrodes with a series of points facing towards the collecting sheets. The gas has an unobstructed path between the electrodes and passes straight through, but the dust particles as they enter the electric field become charged, and are precipitated on to the collectors to which they adhere, the dust gradually building up in this way to a depth of 2 in. or 3 in. Rapping hammers free the collectors from dust. These hammers are operated once every four hours, the dust falling into the hoppers below. The crude gas on an average carries about 5 grammes of dust per cubic metre, though this varies considerably according to the class of iron being made and the ore smelted.

This plant has been in operation for practically four years ; it is at present operating with only one furnace in blast. The dust collected amounts to about 30 tons a week, but this does not give a good idea of the enormous bulk of the dust, as it is extremely fine fume.

The electrical transformer equipment of this plant, which consists of four 25 K.V.A. standard sets, is housed in a building above the precipitator chambers, and each of the four units supplies current to two chambers. The whole plant can be operated by one attendant, apart from the handling of the dust. The dust is removed on day shift only, the hoppers being large enough to take care of a twenty-four hour collection. The removal of the dust from the blast furnace gas enables the gas to be burnt efficiently under boilers and stoves ; approximately only two-thirds of the gas is required for the same heating when the dust has been removed. When clean gas is used there is considerable saving in coke required for the furnace, which often amounts to 2 or 3 cwt. of coke per ton of pig smelted ; in many cases the output of the furnaces is substantially increased, owing to higher blast temperatures and pressures and more uniform working.

The plate type of treater is not always suitable for gas conditions, and a pipe treater is sometimes more convenient. Fig. 7 shows such a pipe treater in course of erection. The installation shown has been in operation for over two years ; there are six units, each unit consisting of 64 pipes. The gases in this installation enter through a 9 ft.-diameter main into the upper part of the hoppers below the pipes, then pass upwards through the pipes into the top headers, and away into the clean main, also 9 ft. diameter. This installation is cleaning the gas from three blast furnaces and has given remarkably good results.

[Table overleaf

Mr. P. E. LANDOLT, Chemical Engineer, Research Corporation, U.S.A., in a Paper gives the following Table of Data.

	Metallurgical Dust and Fume (Large Installation).	Metallurgical Fume.	H₂SO₄ and Precious Metals.	H₂SO₄ Mist.	Precious Metal Fume and Dust.	Metallurgical Fume and Dust.	Air Cleaning (Gold Dust).	Metallurgical Dust.	Metallurgical Dust.	Pyrites Burner Dust.	Producer Gas (Tar).
Investment per cubic feet gas treated	$0.50	$2.38	$3.75	$3.75	$2.85	$2.50	$0.95	$1.50	$0.67	$3.75	$3.00
Cleaning cost per 100,000 cubic feet gas treated per hour	$0.03	$0.13	$0.22	$0.25	$0.23	$0.175	$0.08	$0.08	$0.03	$0.155	$0.15
Temperature of gases treated	250–300° F.	600–700° F.	150° F.	150–200° F.	650° F.	600–700° F.	70° F.	400–500° F.	300–400° F.	900–1000° F.	70° F.
Power consumption, kilowatt-hour per 100,000 cubic feet gas treated	0.3	1.6	2.8	2.0	2.4	2.0	1.0	1.0	0.2	0.1	1.25
Cleaning costs per 100,000 cubic feet gas treated at 70° Fahr.	$0.05	$0.26	$0.25	$0.30	$0.48	$0.35	$0.08	$0.14	$0.04	$0.50	$0.15
Weight of gas treated in tons per hour, approximately	4000	50	26	8.4	8.0	14.4	53	66	270	18	..
Efficiency of removal of suspended matter	85–90%	98+%	99+%	99+%	99+%	97+%	95+%	95%	91%	99%	99%
Weight of precipitate in lb. (average) per 100,000 cubic feet, approx.	6	16	5–15	125	3	8–10	1.0	40–50	..	7.5	15

NOTE.—Gas volumes taken at temperature of treatment, except where otherwise specified.

PLATE I.

A small experimental pipe precipitator,
with electrical arrangement on top.

Fig. 2.
One of the transformers used.

PLATE II.

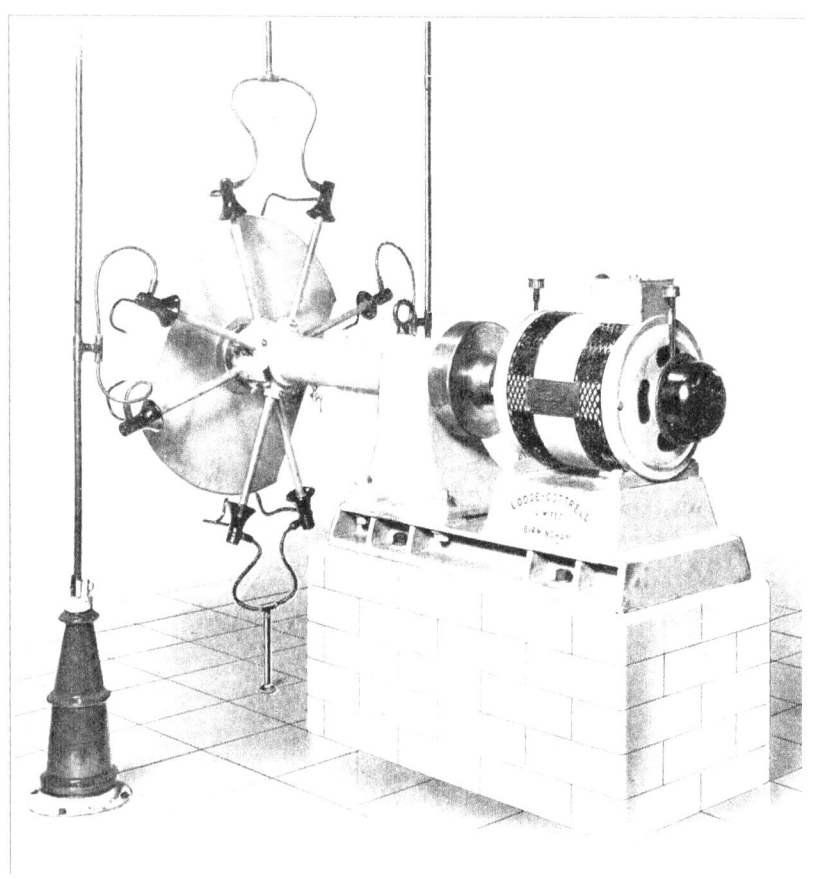

Fig. 3.

A rectifier driven by synchronous motor.

PLATE III.

Fig. 4.

High-tension control room.

PLATE IV.

FIG. 5.

A precipitation chamber for dealing with hot gases from a pyrites-burning
furnace, with hoppers below for dust removal by trucks.

FIG 6.

Fairly large installation of plate type, for cleaning blast furnace gas, erected
during the war. Transformer house above depositing chamber.

PLATE V.

Fig. 7.

A large installation shown in course of erection, with electric control chamber above tube depositors in the middle, and hoppers and railway below.